Structural Design & Drawing

ELEMENTARY Structural Design & Drawing

Volume 1

Dr. D. KRISHNAMURTHY

Ph. D. (Glas), ME (Cal) BE (Hons);
MIE (Inc). MACI, MIABSE (Eng)
Professor of Civil Engineering
M. A. College of Technology, Bhopal
(Presently Professor of Civil Engineering
Sulemaniah University
Sulemaniah)

CBS Publishers & Distributors Pvt. Ltd.

New Delhi • Bengaluru • Chennai • Kochi • Kolkata • Mumbai
Hyderabad • Uttarakhand • Nagpur • Patna • Pune • Jharkhand

ISBN: 81-239-1102-5

First Edition: 1985
Reprint: 1992, 1999, 2003, 2006, 2008, 2009,
2010, 2011, 2012, 2013, 2015, 2018

Published by **Satish Kumar Jain** and produced by **Varun Jain** for
CBS Publishers & Distributors Pvt. Ltd.,
4819/XI Prahlad Street, 24 Ansari Road, Daryaganj, New Delhi - 110002
delhi@cbspd.com, cbspubs@airtelmail.in • www.cbspd.com
Ph.: 23289259, 23266861, 23266867 • Fax: 011-23243014

Corporate Office: 204 FIE, Industrial Area, Patparganj, Delhi - 110 092
Ph: 49344934 • Fax: 011-49344935
E-mail: publishing@cbspd.com • publicity@cbspd.com

Branches:

- ***Bengaluru:*** 2975, 17th Cross, K.R. Road, Bansankari 2nd Stage, Bengaluru - 70 • Ph: +91-80-26771678/79 • Fax: +91-80-26771680 E-mail: cbsbng@gmail.com, bangalore@cbspd.com
- ***Chennai:*** No. 7, Subbaraya Street, Shenoy Nagar, Chennai - 600030 Ph: +91-44-26681266, 26680620 • Fax: +91-44-42032115 E-mail: chennai@cbspd.com
- ***Kochi:*** Ashana House, 39/1904, A.M. Thomas Road, Valanjambalam, Ernakulum, Kochi • Ph: +91-484-4059061-65 Fax: +91-484-4059065 • E-mail: cochin@cbspd.com
- ***Kolkata:*** 6-B, Ground Floor, Rameshwar Shaw Road, Kolkata - 700014 Ph: +91-33-22891126/7/8 • E-mail: kolkata@cbspd.com
- ***Mumbai:*** 83-C, Dr. E. Moses Road, Worli, Mumbai - 400018 Ph: +91-9833017933, 022-24902340/41 • E-mail: mumbai@cbspd.com

Representatives:

- Hyderabad: 0-9885175004
- Patna: 0-9334159340
- Jharkhand: 0-9811541605
- Nagpur: 0-9021734563
- Pune: 0-9623451994
- Uttarakhand: 0-9716462459

Printed at:
J.S. Offset Printers, Delhi (India)

Preface

Structural Engineering is a branch of Civil Engineering which has to deal with the design of buildings, bridges, water towers, retaining walls, silos, bunkers, chimneys and other types of Civil Engineering Structures.

Structural design of any structure may be divided into !—

(*i*) Assessment and distribution of loads which the structural component is required to support or subjected to.

(*ii*) Proportioning of trial section of the components or the appropriate arrangement of structural elements.

(*iii*) Computations of bending moments, shear forces and direct forces due to critical combinations of loads to which the structural member is subjected or is required to carry.

(*iv*) Selection of proper size and shape of the member to resist the forces safely keeping in view the economy in the design.

(*v*) Preparation of lay out of the structure and the finished working drawings with all dimensions for all members which will be required for executing the construction of the structure at site or for fabrication of the structure in the factory.

Students taking courses on structural design have usually no great difficulty in computations involved in the structural design, but they often find that the available text books afford very little guidance in the actual sequence in the design, appropriate choice of section on the basis of strength and economy and the detailed dimensioned drawings.

The author has endeavoured to provide these needs to the students, structural designers handling designs using concrete and steel as materials and also to provide a detailed guide for young engineers and architects engaged in consultancy services.

In the process of making a complete design special attention has been devoted to giving full details of each step. Each design is provided with fully dimensioned working drawings. Designs of all kinds of practical problems that the designer normally comes across are presented using a standard sequence along with the fully dimensioned drawings.

This volume deals with designs with detailed drawings fully dimensioned on practical problems on retaining walls, domed roofs, water tanks on ground, underground, overhead and reservoirs, silos, bunkers, slab and *T*-beam bridges. This book is intended to serve as a text book to civil engineering students of final year, degree course and to those appearing for Section 'B' of the professional examinations and as a reference book for practising engineers in public works, public health departments and structural designers in consulting firms.

AUTHOR

CONTENTS

Elementary Structural Design and Drawing

PART ONE *Pages*

RCC STRUCTURES

I. DESIGN OF SLABS

IV STAIRCASES

Elementary Structural Design and Drawing

PART TWO

STEEL STRUCTURES

V. DESIGN OF STRUCTURAL CONNECTIONS

VI DESIGN OF BEAMS

IX. DESIGN OF RIVETED PLATE GIRDERS

Notations For Drawings

(i) G_1—MR—10—185
G_1—Grade 1 Steel
MR—Mild Steel round bars
10—10 mm diameter
185—185 mm spacing a/c

(ii) RT—10—200
RT—Ribbed
10—10 mm diameter
200—200 mm spacing c/c

(iii) 9—G_1—MR—25
9—9 Numbers of bars
G_1—Grade I Steel
MR—Mild Steel round bars
25—25 mm diameter

(iv) 9—RT—20
9—9 Number of bars
RT—Ribbed Torsteel
20—20 mm diameter

I. Design of Slabs

1

Slabs Spanning in one Direction—General

1.1 Definition

A concrete slab is said to be spanning in one direction when it is supported on two opposite sides only. A slab, when supported on all four sides and if the length of the slab exceeds three times its width, is also said to be spanning in one direction.

1.2 Slab Depth

The maximum value of span/depth ratio = 30

i.e., $$d = \frac{\text{Span}}{30}$$

1.3 Effective Span

It is the smaller of the following :

(*i*) Distance between centre to centre of the supports.

(*ii*) Clear distance between the supports + the effective depth of the slab

1.4 Loads (Dead Loads)

The self-weight of the slab, partition walls and finishes etc. shall be taken as the dead load on the slab. Loads due to partition walls shall be assessed on the basis of actual constructional details of the proposed partitions and their location or position. The loads assumed shall be included in the dead load for the design of the floors and the supporting structure.

When the loads due to partition walls cannot be assumed before hand, a uniformly distributed load per square metre of not less than 33⅓% of the weight per metre length of finished partition on the entire floor are subjected to a minimum u.d.l. of 100 kg/m² shall be taken as dead load.

1.5 Live Loads (Floors)

Loading Class	*Types of floors*	*Minimum Live load* kg/m²
200	Floors in dwelling houses, tenements, hospital wards, bedrooms and private sitting rooms in Hostels and dormitories.	200

250	Office floors other than entrance hall floors of light work rooms.	250 400
300	Floors of banking halls, office entrance halls and reading rooms.	300
400	Shop floors used for the display and sale of merchandise, floors of work rooms, floors of class rooms, in restaurants, circulation space in machinery halls, power stations etc. where not occupied by plant or equipment.	400
500	Floors of ware houses, workshops, factories and other buildings or parts of building of similar category for light weight loads, office floors for storage and filling purposes, floors of place of assembly without fixed seating, public rooms in hotels, dance halls, waiting halls etc.	500
750	Floors of ware houses, workshops, factories and other buildings or part of buildings of similar category for medium weight loads.	750
1000	Floors of ware houses, workshops, factories and other buildings, or parts of buildings of similar category for heavy weight loads, floors of book stores and libraries, roofs and pavement lights over basements projecting under the public foot path.	1000

1·6 Garage (Light)

Floors used for garages for vehicles not exceeding 2.5 tonnes gross weight.	
Slabs	400
Beams	250

1·6·1 Garage (Heavy)

Floors used for garages for vehicles not exceeding 4 tonnes gross weight.	750

1·7 Stairs

Stairs, Landings and corridors for class 200, but not liable to over-crowding.	300
Stairs, Landings and corridors for class 200 loading but liable to over-crowding and for all other classes.	500

1·8 Balcony

Balconies not liable to over-crowding for class 200 Loading.	300

For all other classes	500
Balconies liable to over-crowding	500

1·9 Roofs

Type of roof	Live load in Plan kg/m²
Flat, sloping or curved roof with slopes upto and including 10 degrees	
(*a*) Access provided	150
(*b*) Access not provided, except for maintenance.	75
Sloping roof with slope greater than 10°	75 less one kg/m² for every increase in slope over 10 degrees upto and including 20° and 2 kg/m² for every degree increase in slope over 20°

1·10 Dead Loads of Materials

Reinforced concrete	=2400 kg/m³
Cement plaster 5 mm thick	=10 kg/sq m
10 mm thick	=20 kg/sq m
Mastic asphalt 10 mm thick	=22 kg/sq m
Asphalt flooring 10 mm thick	=22 kg/sq m
I.P.S. Floor 25 mm thick	=60 kg/sq m
Concrete tile flooring 25 mm thick	=50 kg/sq m
Terrazo 10 mm thick	=20 kg/sq m
Brick walls 100 mm thick	=195 kg/sq m
Rubble masonary	=2080 kg/m³

1·11 Bending Moments

The bending moments shall be calculated for the effective spans and for all loading thereon.

Slabs carrying concentrated load should be designed to resist the maximum bending moment by the loading system. The bending moment so produced shall be assumed to be resisted by an effective width of slab, measured parallel to the supporting edges as under :

(*i*) When a single concentrated load acts.

B_e=the effective width of slab shall not exceed the value given by the expression, provided that it shall not exceed the actual width of the slab.

$$B_e = kx\left(1-\frac{x}{L_e}\right)+b_e$$

where

x=distance of the concentrated load from the nearer support.

b_e=width of cöntact area of the concentrated load measured parallel to the supported edge.

l_e=Effective span.

k=a constant depending upon the ratio of the width of the slab l' to the effective span l_e.

$\frac{l'}{l_e}$	0·1	0·2	0·3	0·4	0·5	0·6	0·7	0·8	0·9
k	0·4	0·8	1·16	1·48	1·72	1·96	2·12	2·24	2·3

For $\frac{l'}{l_e}=1$ and above, $k=2{\cdot}48$.

(*ii*) If two or more concentrated loads acting in a line in the direction of span the bending moment per metre width of slab shall be calculated separately for each load according to its appropriate effective width of slab as calculated in (*i*) and added together.

(*iii*) If two or more loads not acting in a line in the direction of the span and the effective width of slab for one load does not overlap the effective width of slab for another load, both calculated as in (*i*) then the slab for each load can be designed separately.

If the effective width of slab for one load overlaps the effective width of slab for an adjacent load the overlapping portion of the slab shall be designed for the combined effect of the two loads.

1·12 Allowable Stresses

Concrete

Concrete grade	Bending kg/cm²	Direct compres-sion kg/cm²	Shear kg/cm²	Bond stresses kg/cm²		Modular ratio
				Average	Local	
M 100	30	25	3	4	7	31
M 150	50	40	5	6	10	19
M 200	70	50	7	8	13	13
M 250	85	60	8	9	15	11
M 300	100	80	9	10	17	9
M 350	115	90	10	11	18	8
M 400	130	100	11	12	19	7

1·12·1 Steel

For grade—1 steel

Tensile stress in bars upto and including 40 mm diameter
=1400 kg/cm²

bars over 40 mm diameter=1300 kg/cm²

1·13 Minimum Reinforcement

The reinforcement in either direction shall not be less than 0·15 percent of the gross-sectional area of the concrete.

1·14 Cover

(*i*) A minimum of 13 mm or not less than the diameter of the reinforcement whichever is greater.

(*ii*) At each end of reinforcing bar not less than 25 mm or not less than twice the diameter of such bar.

1·15 Spacing Mainsteel

The spacing of the bars of main tensile reinforcement in solid slabs shall not be more than three times the effective depth or 600 mm whichever is small.

1·15·1 Secondary Steel

The spacing of the distribution or secondary steel provided for shrinkage and temperature shall not be more than five times the effective depth of such slabs or 600 mm whichever is smaller.

1·16 Shear and Bond Stresses

The shear stress shall not exceed the value given by the expression.

$$q_s = \frac{Q}{bjd}$$

and the local bond stress

$$\sigma_{bl} = \frac{Q}{\Sigma O j d}$$

1·17 Design Coefficients

σ_{cb} kg/cm²	m	σ_{st} kg/cm²	R kg/cm²	n	j
50	19	1000	10·22	0·488	0·838
		1400	8·74	0·404	0·865
		2300	6·59	0·292	0·903
70	13	1000	14·01	0·476	0·841
		1400	11·98	0·393	0·868
		2300	8·97	0·283	0·906

m=modular ratio$=\dfrac{2800}{3\sigma_{cb}}$

n=depth of NA coefficient$=\dfrac{1}{1+\dfrac{\sigma_{st}}{m\sigma_{cb}}}$

$j=1-\dfrac{n}{3}$=Lever arm coefficient

$R=\dfrac{1}{2}\sigma_{cb}\times n\times j$

$M=R\,bd^2{}_e$=Resisting moment

d_e=effective depth$=\sqrt{\dfrac{M}{R_b}}$

A_{st}=Amount of steel$=\dfrac{M}{\sigma_{st}\times jd_e}$

2

Simply Supported Corridor Slab Spanning in one Direction

2·1 Data

Clear Span=2·1 m
Supported on brickwalls of 40 cm on one side and 20 cm brick wall on the other.
Loading as per school buildings
Materials : Concrete M 150, Grade—I steel.

2·2 Relevant Codes

IS—456, IS—875

2·2·1 Allowable Stresses

$C=50$ kg/cm^2 $\quad t=1400$ kg/cm^2
$q_s=5$ kg/cm^2 $\quad m=19$
$\quad R=8{\cdot}74$
$q_b=10$ kg/cm^2 $\quad a={\cdot}865\, d_e$

2·3 Depth of Slab

For simply supported slab spanning in one direction

$$d=\frac{\text{Span}}{30}=\frac{210}{30}=7 \text{ cm}$$

assume $d=10$ cm, $\quad d_e=8$ cm

2·4 Effective Span

Least of

(*i*) Centre to centre of supports$=2{\cdot}1+{\cdot}20+{\cdot}10$
$=2{\cdot}4$ m

(*ii*) clear span+effective depth $=2{\cdot}1+{\cdot}08=2{\cdot}18$ m
$l=2{\cdot}18$ m

2·5. Loads

Live load on floors of schools=400 kg/m^2
Dead load of slab 10 cm =240 kg/m^2
Floor Finish (2·5 cm IPS) = 60 kg/m^2
Total load=w=700 kg/m^2

2·6 Bending Moments

$$M=\frac{wl^2}{8}=\frac{700\times 2{\cdot}18^2}{8}=414 \text{ mkg}$$

2·7 Shear Force

$$S=\frac{wl}{2}=\frac{700\times2{\cdot}18}{2}=765 \text{ kg}$$

2·8 Effective Depth Needed

$$d_e=\sqrt{\frac{M}{b\times R}}=\sqrt{\frac{414\times100}{100\times8{\cdot}74}}$$

$$=\sqrt{47}=6{\cdot}85 \text{ cm}$$

Use $d=10$ cm, $d_e=8$ cm

2·8·1 Main Steel

$$A_t=\frac{41400}{1400\times{\cdot}865\times8}=4{\cdot}25 \text{ cm}^2$$

Use 10 mm at 18 cm c/c

Max spacing allowed$=3\times d_e=3\times8=24$ cm

2·8·2 Secondary Steel

$$A_s=\frac{0{\cdot}15}{100}\times100\times10=1{\cdot}5 \text{ cm}^2$$

Use 6 mm at 18·5 cm c/c

2·8·3 Shear Stress

$$q_s=\frac{S}{b\times a}=\frac{765}{100\times{\cdot}865\times8}$$

$$=1{\cdot}09 \text{ kg/cm}^2 < 5 \text{ kg/cm}^2$$

2·8·4 Bond Stress

$$q_b=\frac{S}{a\Sigma o}$$

At the supports alternate bars are bent up.

$$\text{number of bars}=\tfrac{1}{2}\left(\frac{100}{18}+1\right)=\frac{6{\cdot}6}{2}=3{\cdot}3$$

$$q_b=\frac{765}{{\cdot}865\times8\times\pi\times1\times3{\cdot}3}$$

$$=10{\cdot}65 \text{ kg/cm}^2$$

$$>10 \text{ kg/cm}^2$$

Use 6 mm additional bars at support at 18 cm c/c at least 60 cm long.

3

Simply Supported Corridor Slab Spanning in one Direction

3·1 Data

Clear span=2·1 m
Supported on brick walls of 40 cm on one side and 20 cm brick wall on the other
Materials : Concrete M 150, Steel—Ribbed Torsteel

3·2 Relevant Codes

IS—456, IS—875

3·2·1 Allowable Stresses

$C=50$ kg/cm^2 $\quad t=2300$ kg/cm^2
$q_s=5$ kg/cm^2 $\quad m=19$
$q_b=14$ kg/cm^2 $\quad a=0{\cdot}903\, d_e$
$R=6{\cdot}59$

3·3 Depth of Slab

For simply supported slab spanning in one direction.

$$d=\frac{\text{span}}{30}=\frac{210}{30}=7 \text{ cm}$$

Assume $d=10$ cm, $d_e=8$ cm

3·4 Effective Span

Least of

(*i*) Centre to centre of supports$=2{\cdot}1+{\cdot}20+{\cdot}10$
$=2{\cdot}4$ m

(*ii*) Clear span+effective depth$=2{\cdot}1+{\cdot}08=2{\cdot}18$ m
$l=2{\cdot}18$ m

3·5 Loads

Live load on floor of schools	=400 kg/m^2
Dead load of slab 10 cm	=240 kg/m^2
Floor Finish etc	=60 kg/m^2
Total load=w	=700 kg/m^2

3·6 Bending Moment

$$M=\frac{wl^2}{8}=\frac{700\times 2{\cdot}18^2}{8}=414 \text{ mkg}$$

3·7 Shear Force

$$S=\frac{wl}{2}=\frac{700\times 2{\cdot}18}{2}=765 \text{ kg}$$

3·8 Effective depth needed

$$d_e = \sqrt{\frac{414 \times 100}{100 \times 6{\cdot}59}} = \sqrt{63}$$

$$= 7{\cdot}9 \text{ cm}$$

Use $d = 10$ cm, $d_e = 8$ cm

3·8·1 Main Steel

$$A_t = \frac{41400}{2300 \times {\cdot}903 \times 8} = 2{\cdot}49 \text{ cm}^2$$

Use 8 mm—20 cm c/c
max. spacing allowed$= 3 \times d_e = 3 \times 8 = 24$ cm c/c

3·8·2 Secondary Steel

$$A_s = \frac{0{\cdot}15 \times 10 \times 100}{100} = 1{\cdot}5 \text{ cm}^2$$

Use 6 mm—18 cm c/c

3·8·3 Shear Stress

$$q_s = \frac{S}{b \times a} = \frac{765}{100 \times {\cdot}903 \times 8}$$

$$= 1{\cdot}06 \text{ kg/cm}^2 < 5 \text{ kg/cm}^2$$

3·8·4 Bond Stress

$$q_b = \frac{S}{a \Sigma o}$$

At the support alternate bars are bent

$$\text{number of bars} = \tfrac{1}{2}\left(\frac{100}{20} + 1\right) = 3$$

$$q_b = \frac{765}{{\cdot}903 \times 8 \times \pi \times {\cdot}8 \times 3}$$

$$= 14 \text{ kg/cm}^2$$

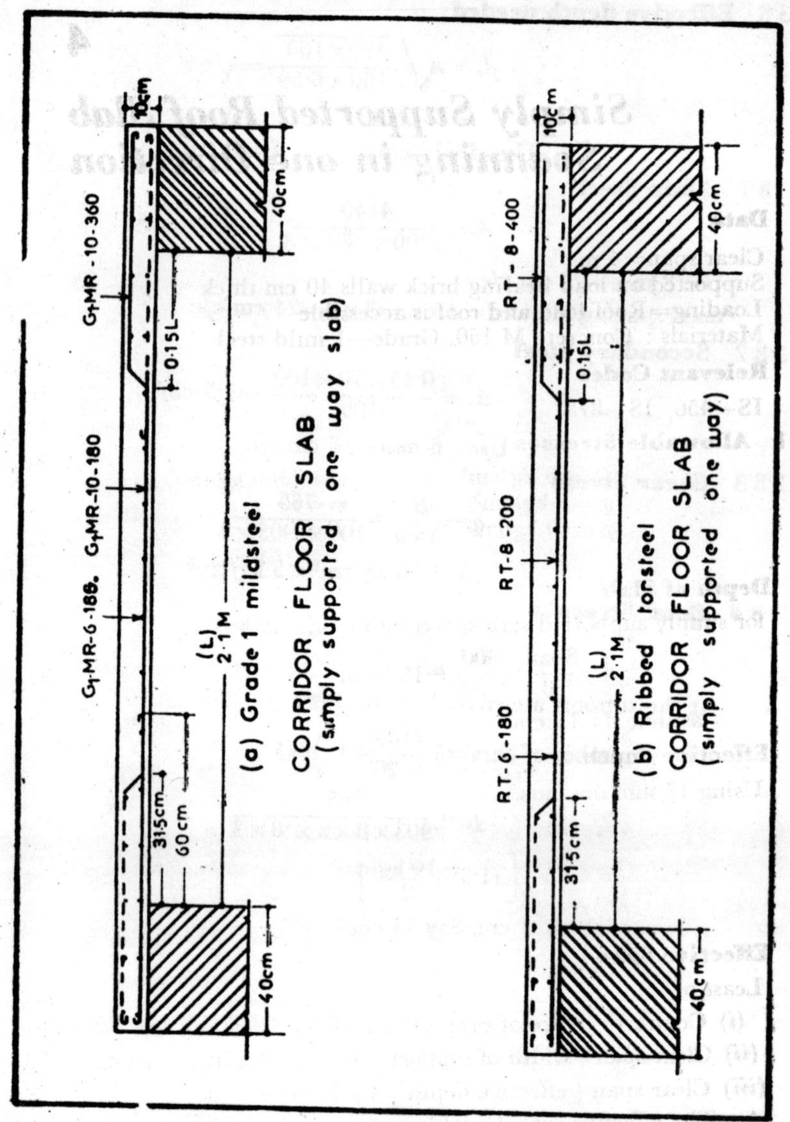

(a) Grade 1 mildsteel
CORRIDOR FLOOR SLAB
(simply supported one way slab)

(b) Ribbed torsteel
CORRIDOR FLOOR SLAB
(simply supported one way)

4

Simply Supported Roof Slab Spanning in one Direction

4·1 Data

Clear span=4 m
Supported on load bearing brick walls 40 cm thick
Loading=Roof load and roof is accessible
Materials : Concrete M 150, Grade—1 mild steel

4·2 Relevant Codes

IS—456, IS—875

4·2·1 Allowable Stresses

$C=50$ kg/cm^2	$t=1400$ kg/cm^2
$q_s=5$ kg/cm^2	$m=19$
$q_b=10$ kg/cm^2	$R=8{\cdot}74$
	$a=0{\cdot}865\ d_e$

4·3 Depth of Slab

for simply supported slab spanning in one direction

$$d=\frac{\text{Span}}{30}=\frac{400}{30}=13{\cdot}5 \text{ cm}$$

assume $d=15$ cm

4·4 Effective depth

Using 12 mm dia bars.

$$d_e=15-(\text{cover}+\tfrac{1}{2}\text{ diameter of the bar})$$

$$=15-\left[(1{\cdot}3)+\frac{(1{\cdot}2)}{2}\right]$$

$$=13{\cdot}1 \text{ cm, Say } 13 \text{ cm}$$

4·5 Effective Span

Least of,

(*i*) Centre to centre of supports=4+0·4=4·4 m
(*ii*) Clear span+width of support=4+0·4=4·4 m
(*iii*) Clear span+effective depth=4+0·13=4·13 m
∴ The effective span l=4·13 m

4·6 Loads

Live load on accessible roofs=150 kg/m^2

Dead load of slab (15 cm) $=\dfrac{15\times1\times2400}{100}=360$ kg/m^2

Roof finish, bituminous felt for water proofing=3 kg/m²

Total load w=513 kgm²

Say w=515 kg/m²

4·7 Bending Moment

$$M=\frac{wl^2}{8}=\frac{515\times 4{\cdot}13^2}{8}=1090 \text{ mkg}$$

4·8 Shear Force

$$S=\frac{wl}{2}=\frac{515\times 4{\cdot}13}{2}=1060 \text{ kg}$$

4·9 Effective Depth Required

$$d_e=\sqrt{\frac{M}{b\times R}}=\sqrt{\frac{1090\times 100}{100\times 8{\cdot}74}}=\sqrt{124{\cdot}2}=11{\cdot}1 \text{ cm}$$

Use d=13 cm, d_e=11 cm

4·9·1 Main Steel

$$A_t=\frac{109000}{1400\times 0{\cdot}865\times 11}=8{\cdot}25 \text{ cm}^2$$

Use 12 mm at 13·5 cm c/c

Maximum spacing allowed=3×effective depth

=3×11=33 cm

or 60 cm whichever is less.

4·9·2 Secondary Steel

$$A_s=\frac{0{\cdot}15}{100}\times 100\times 13=1{\cdot}95 \text{ cm}^2$$

Use 6 mm at 14·5 cm c/c

Maximum spacing allowed=5×11=55 cm

4·9·3 Shear Stress

$$q_s=\frac{S}{b\times a}$$

$$q_s=\frac{1060}{100\times{\cdot}865\times 11}=1{\cdot}11 \text{ kg/cm}^2<5 \text{ kg/cm}^2$$

4·9·4 Bond Stress

At the support alternate bars are bent up

∴ number of bars $=\frac{1}{2}\left(\frac{100}{13{\cdot}5}+1\right)=4{\cdot}2$

$$q_b=\frac{S}{a\,\Sigma o}=\frac{1060}{{\cdot}865\times 11\times\pi\times 1{\cdot}2\times 4{\cdot}2}$$

$$=7{\cdot}05 \text{ kg/cm}^2<10 \text{ kg/cm}^2$$

5

Simply Supported Roof Slab Spanning in one Direction

5·1 Data

Clear span=4 m
Supported on load bearing brick walls 40 cm thick
Loading—Roof load and roof is accessible.
Materials : Ribbed Torsteel, Concrete M 150

5·2 Relevant Codes

IS—456, IS—875

5·2·1 Allowance Stresses

$C=50$ kg/cm² $\quad t=2300$ kg/cm²
$q_s=5$ kg/cm² $\quad a=0{\cdot}903\ d_e$
$q_b=14$ kg/cm² $\quad R=6{\cdot}59$

5·3 Depth of Slab

For simply supported slab spanning in one direction

$$d=\frac{\text{Span}}{30}=\frac{400}{30}=13{\cdot}5 \text{ cm}$$

assume $d=15$ cm

5·4 Effective Depth

Using 12 mm dia bars

$$d_e=15-(\text{cover}+\tfrac{1}{2}\text{ dia of bar})$$
$$=15-\left(1{\cdot}3+\frac{1{\cdot}2}{2}\right)$$
$$=13{\cdot}1 \text{ cm Say } 13 \text{ cm}$$

5·5 Effective Span

Least of

(*i*) Centre to centre of supports $=4+0{\cdot}4=4{\cdot}4$ m
(*ii*) Clear span+width of supports $=4+0{\cdot}4=4{\cdot}4$ m
(*iii*) Clear span+effective depth $=4+0{\cdot}13=4{\cdot}13$ m

∴ The effective span $l=4{\cdot}13$ m

5·6 Loads

Live load on accessible roofs=150 kg/m²

Dead load of slab (15 cm) $=\dfrac{15\times 1\times 2400}{100}=360$ kg/m²

Roof finish, Bituminous felt for water proofing=3 kg/m²

Total load $w=513$ kg/m²
Say $=515$ kg/m²

5·6·1 Bending Moment

$$M=\frac{wl^2}{8}=\frac{515\times4\cdot13^2}{8}=1090 \text{ mkg}$$

5·6·2 Shear Force

$$S=\frac{wl}{2}=\frac{515\times4\cdot13}{2}=1060 \text{ kg}$$

5·6·3 Effective Depth required

$$d_e=\sqrt{\frac{M}{b\times R}}=\sqrt{\frac{1090\times100}{100\times6\cdot59}}$$

$$=\sqrt{166} \quad =12\cdot9 \text{ cm}$$

Use $d_e=13$ cm $\quad d=15$ cm

5·6·4 Main Steel

$$A_t=\frac{109000}{2300\times\cdot903\times13}=4\cdot01 \text{ cm}^2$$

Use 12 mm at 28 cm c/c
Maximum spacing allowed$=3\times d_e=3\times13=39$ cm
Hence spacing adopted is with in the permissible limits.

5·6·5 Secondary Steel

$$A_s=\frac{0\cdot15\times100\times}{100}=2\cdot25 \text{ cm}^2$$

Use 8 mm ϕ at 22 cm c/c
Maximum spacing allowed$=5\times13=65$ cm

5·7 Shear Stress

$$q_s=\frac{S}{b\times a}$$

$$=\frac{1060}{100\times\cdot903\times13}=0\cdot905 \text{ kg/cm}^2 < 5 \text{ kg/cm}^2$$

5·8 Bond Stress

At the supports alternate bars are bent up

Number of bars$=\frac{1}{2}\left(\frac{100}{28}+1\right)=2\cdot28$

$$q_b=\frac{S}{a\times\Sigma o}$$

$$=\frac{1060}{0\cdot903\times13\times\pi\times1\cdot2\times2\cdot28}$$

$$=10\cdot4 \text{ kg/cm}^2$$

$$< 14 \text{ kg/cm}^2$$

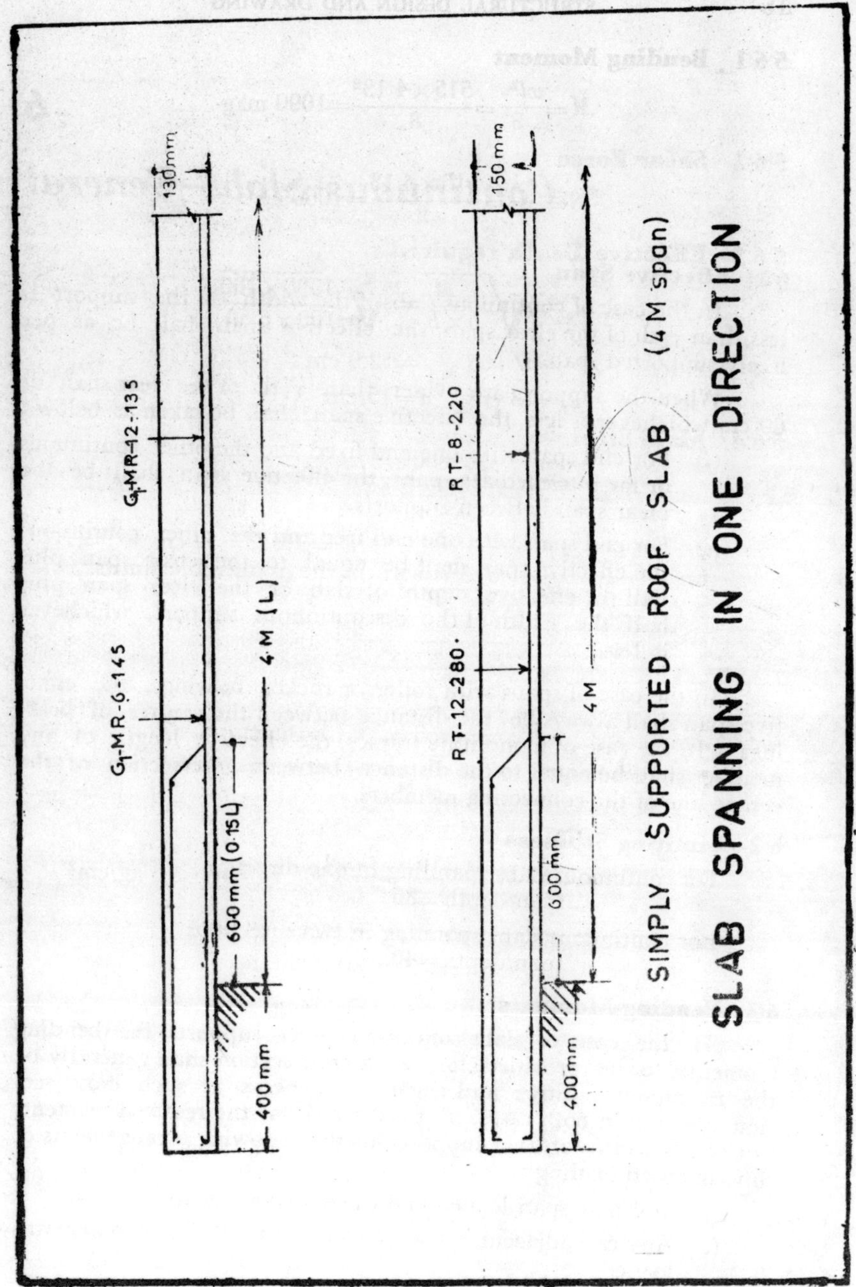
130mm
G1-MR-12-135
G1-MR-6-145
600mm (0.15L)
4 M (L)
400mm
150 mm
RT-8-220
RT-12-280
4 M
600mm
400mm
SIMPLY SUPPORTED ROOF SLAB (4 M span)
SLAB SPANNING IN ONE DIRECTION

6

Continuous Slabs–General

6·1 Effective Span

In the case of continuons slabs if the width of the support is less than $\frac{1}{12}$th of the clear span, the effective span shall be as per freely supported span.

When the supports are wider than $\frac{1}{12}$th of the clear span or 60 cm whichever is less, the effective span, shall be taken as below :

(*i*) For end span with one end fixed and the other continuous or for intermediate spans, the effective span shall be the clear span between supports.

(*ii*) For end span with one end free and the other continuous the effective span shall be equal to the clear span plus half the effective depth of slab or the clear span plus half the width of the discontinuous support, whichever is less.

In the case of spans with roller or rocker bearings, the effective span shall always be the distance between the centres of bearings. In the case of monolithic frames, the effective length of any member shall be equal to the distance between intersection of the centre line of the connecting members.

6·2 Limiting Stiffness

For continuous slabs spanning in one direction,

span/depth=35.

For continuous slabs spanning in two directions,

span/depth=40.

6·3 Bending Moments

In the case of slabs continuous over supports the bending moments to be provided for, at a cross section shall generally be the maximum positive and negative moments at such cross section allowing in both cases, if so desired for the reduced moments due to the width of the supports for the following arrangements of superimposed loading :

(*i*) alternate span loaded and other spans unloaded

(*ii*) Any two adjacent spans loaded and all other spans unloaded.

The bending moments in uniformly loaded beams and slabs continuous over three or more approximately equal spans may be assumed to have the following values :

Description	Near middle of end span	At middle of interior span	At support next to the end support	At other interior supports
Moment due to Dead load	$+\frac{W_d l}{12}$	$+\frac{W_d l}{24}$	$-\frac{W_d l}{10}$	$-\frac{W_d l}{12}$
Moment due to superimposed load	$+\frac{W_s l}{10}$	$+\frac{W_s l}{12}$	$-\frac{W_s l}{9}$	$-\frac{W_s l}{9}$

where W_d=total dead load per span
W_s=total superimposed uniformly distributed load per span
l=Effective span.

6·4 Shear Forces

Corresponding to the bending moments given above the shearing forces at the supports may be assumed to have the following values :

Description	At exterior face of support next to the end support	At faces of all other supports
Shearing force due to dead load	0·6 W_d	0·5 W_d
Shearing force due to superimposed load	0·5 W_s	0·5 W_s

6·5 Concentrated Loads

For continuous slabs carrying concentrated loads, the effective width of the slab shall be calculated as in the case of simply supported slabs carrying concentrated loads using the values of k given in the Table.

$\frac{l'}{l_e}$ =	0·1	0·2	0·3	0·4	0·5	0·6	0·7	0·8	0·9	1 and above
k=	0·4	0·8	1·16	1·44	1·68	1·84	1·96	2·08	2·16	2·24

Two—Way Continuous Slabs

The design shall be carried out as per requirements given under two way slabs.

6·6 Unsymmetrical Loading

When the loads on the various spans are not the same, it is necessary to derive the bending moment diagram from the actual loading conditions. This can be done by the theorem of three moments. A quicker method is to do graphically by the use of the characteristic point method or by moment distribution method.

7

Class Room Floor Slab Continuous and Spanning in one Direction–General

7·1 Data

Class room size = 7 m × 8·25 m
Slab is continuous over *T*—beam ribs
Centre to centre of *T*—beam rib = 2·75 m
Rib width = 25 cm
Loading = Class room floor
Design—Interior span
Materials : Grade 1—Steel, Concrete M 150

7·2 Relevant Codes

IS—456, IS—875

7·2·1 Allowable Stresses

$C = 50$ kg/cm² $\quad t = 1400$ kg/cm²
$q_s = 5$ kg/cm² $\quad m = 19$
$q_b = 10$ kg/cm² $\quad a = \cdot 865\ d_e$

7·3 Type

Slab is supported on two parallel ribs of the *T*-beams and

$$\frac{L}{B} = \frac{7}{2\cdot 75} = 2\cdot 55 > 2.$$

The slab is continuous on supports in one direction only.

7·4 Depth of Slab

For continuous slab spanning in one direction

$$d = \frac{\text{Span}}{35} = \frac{275}{35} = 7\cdot 85 \text{ cm}$$

Assume overall depth = 10 cm

7·4·1 Effective Depth

Using 10 mm bars

$d_e = 10 - $(clear cover + $\frac{1}{2}$ the diameter of the bar)
$= 10 - (1\cdot 3 + 0\cdot 5)$
$= 8\cdot 2$ cm

Use $d_e = 8$ cm $\quad d = 10$ cm

7·4·2 Effective Span

Width of support = 25 cm
1/12 clear span = 1/12 × 250 = 20·8 cm
Width of support is wider than 1/12 span

For interior span the effective span

$$=\text{clear span between supports}$$
$$l=2{\cdot}5 \text{ m}$$

7·5 Loads

Superimposed live load (floors for schools)

$$w_s=400 \text{ kg/m}^2$$

Dead load of slab (10 cm) $=\dfrac{10}{100}\times 1\times 2400=240$ kg/m²

Floor finish (2·5 cm) IPS $=60$ kg/m²

Total dead load w_d $=300$ kg/m²

7·5·1 Bending Moments

BM (−ve) at the interior support

$$=\frac{w_d l^2}{12}+\frac{w_s l^2}{9}$$
$$=\frac{300\times 2{\cdot}5^2}{12}+\frac{400\times 2{\cdot}5^2}{9}$$
$$=156+277{\cdot}5=433{\cdot}5 \text{ mkg}$$

BM at centre of interior span (+ve)

$$=\frac{w_d l^2}{24}+\frac{w_s l^2}{12}$$
$$=\frac{300\times 2{\cdot}5^2}{24}+\frac{400\times 2{\cdot}5^2}{12}$$
$$=78+206=284 \text{ mkg}$$

The support moment is greater than the moment at centre. Therefore 15% decrease in support moment and 15% increase in central moment is allowed.

7·5·2 Revised Moments

Support moment $=433{\cdot}5-0{\cdot}15\times 433{\cdot}5$

$=433{\cdot}5-65$

$=368{\cdot}5$ mkg

Central moment $=284+65=349$ mkg

Maximum moment $=368{\cdot}5$ mkg

7·5·3 Shear Force

Shear force at interior support

$$S=0{\cdot}5\ W_d+0{\cdot}5\ W_s$$
$$=0.5\times 300\times 2{\cdot}5+0{\cdot}5\times 400\times 2{\cdot}5$$
$$=375+500=875 \text{ kg}$$

7·5·4 Effective Depth Needed

$$d_e=\sqrt{\frac{M}{b\times R}}=\sqrt{\frac{36850}{100\times 8{\cdot}74}}$$
$$=\sqrt{42}=6{\cdot}5 \text{ cm}$$

Use $d_e=8$ cm $\quad d=10$ cm

7·5·5 Main Steel

$$A_t=\frac{M}{t\times a}$$

$$=\frac{36850}{1400\times \cdot 865\times 8}=3\cdot 9 \text{ cm}^2$$

Use 10 mm at 20 cm c/c

Maximum spacing allowed$=3\times d_e=3\times 8=24$ cm

7·5·6 Secondary Steel

$$A_s=\frac{0\cdot 15}{100}\times 10\times 100=1\cdot 5 \text{ cm}^2$$

Use 6 mm at 18·5 cm c/c

Max. spacing allowed$=5\times d_e=5\times 8=40$ cm

7·5·7 Shear Stress

$$q_s=\frac{S}{b\times a}=\frac{875}{100\times \cdot 865\times 8}=1\cdot 27 \text{ kg/cm}^2$$

7·5·8 Bond Stress

$$q_b=\frac{S}{a\Sigma o}$$

At support alternate bars are bent up from either side.

$$\text{Number of bars}=\frac{100}{20}+1=6$$

$$q_b=\frac{875}{\cdot 865\times 8\times \pi\times 6\times 1}=6\cdot 7 \text{ kg/cm}^2$$

$$< 10 \text{ kg/cm}^2$$

8

Class Room Floor Slab Continuous and Spanning in one Direction

8·1 Data

Class room size = 7m × 8·25 m
Slab is continuous over *T*-beam ribs
Centre to centre of *T*-beam ribs = 2·75 m
Rib width = 25 cm
Loading—class room floor
Design interior span
Materials : Ribbed torsteel, Concrete M 150

8·2 Relevant Codes

IS—456, IS—875

8·2·1 Allowable Stresses

$C=50$ kg/cm², $q_s=5$ kg/cm², $q_b=14$ kg/cm²
$t=2300$ kg/cm², $m=19$, $R=6{\cdot}59$ $a={\cdot}903\ d_e$

8·3 Type

Slab is supported on two parallel ribs of

$$T\text{—beams and } \frac{L}{B}=\frac{7}{2{\cdot}75}=2{\cdot}55>2$$

The slab is continuous on supports in one direction only.

8·4 Depth of Slab

For continuous slab spanning in one direction

$$d=\frac{\text{span}}{35}=\frac{275}{35}=7{\cdot}85 \text{ cm}$$

8·4·1 Effective Depth

$$d_e=8 \text{ cm and } d=10 \text{ cm}$$

8·4·2 Effective Span

Width of support = 25 cm
1/12 clear span = 1/12 × 250 = 20·8 cm < 25 cm
Width of support is wider than 1/12 span
For interior span the effective span
= clear span between supports = 2·5 m
$l=2{\cdot}5$ m

8·5 Loads

Superimposed load (floors for schools) $w_s=400$ kg/m²

Dead load of slab (10 cm) $=\frac{10}{100}\times1\times2400=240$ kg/m²

Floor finish (2·5 cm) IPS=60 kg/m^2
Total dead load $=w_d=$300 kg/m^2.

8·5·1 Bending Moments

BM (−ve) at the interior support

$$=\frac{w_d l^2}{12}+\frac{w_s l^2}{9}=\frac{300\times2\cdot5^2}{12}+\frac{400\times2\cdot5^2}{9}$$
$$=156+277\cdot5=433\cdot5 \text{ mkg.}$$

BM at centre of interior span (+ve)

$$=\frac{w_d l^2}{24}+\frac{w_s l^2}{12}=\frac{300\times2\cdot5^2}{24}+\frac{400\times2\cdot5^2}{12}$$
$$=78+206=284 \text{ mkg.}$$

The support moment is greater than the moment at centre. Therefore, 15% decrease in support moment and 15% increase in central moment is allowed.

8·5·2 Revised Moments

Support moment$=433\cdot5-0\cdot15\times433\cdot5$
$=433\cdot5-65=368\cdot5$ m kg.

8·5·3 Shear Force

Shear force at interior support

$$S=0\cdot5W_d+0\cdot5W_s$$
$$=0\cdot5\times300\times2\cdot5+0\cdot5\times400\times2\cdot5=375+500$$
$$=875 \text{ kg.}$$

8·5·4 Effective depth Needed

$$d_e=\sqrt{\frac{M}{b\times R}}=\sqrt{\frac{36850}{100\times6\cdot59}}=\sqrt{56}=7\cdot5 \text{ cm.}$$

Use d_e=8 cm d=10 cm.

8·5·5 Main Steel

$$A_s=\frac{36850}{2300\times\cdot903\times8}=2\cdot02 \text{ cm}^2.$$

Use 8 mm ϕ at 24 cm c/c.
Maximum spacing allowed$=3\times d_e=3\times8=24$ cm.
Therefore adopted spacing is alright.

8·5·6 Secondary Steel

$$A_e=\frac{0\cdot15}{100}\times100\times10=1\cdot5 \text{ cm}^2.$$

Use 6 mm ϕ at 18·5 cm c/c.
Maximum spacing allowed$=5\times8=40$ cm.

8·5·7 Shear Stress

$$q_s=\frac{S}{b\times a}=\frac{875}{100\times\cdot903\times8}=1\cdot27 \text{ kg/cm}^2.$$

8·5·8 Bond Stress

At support alternate bars are bent up from either side.

Number of bars$=\frac{100}{24}+1=5\cdot2$.

$$q_b=\frac{875}{\cdot903\times8\times\pi\times0\cdot8\times5\cdot2}$$
$$=9\cdot3 \text{ kg/cm}^2<14 \text{ kg/cm}^2.$$

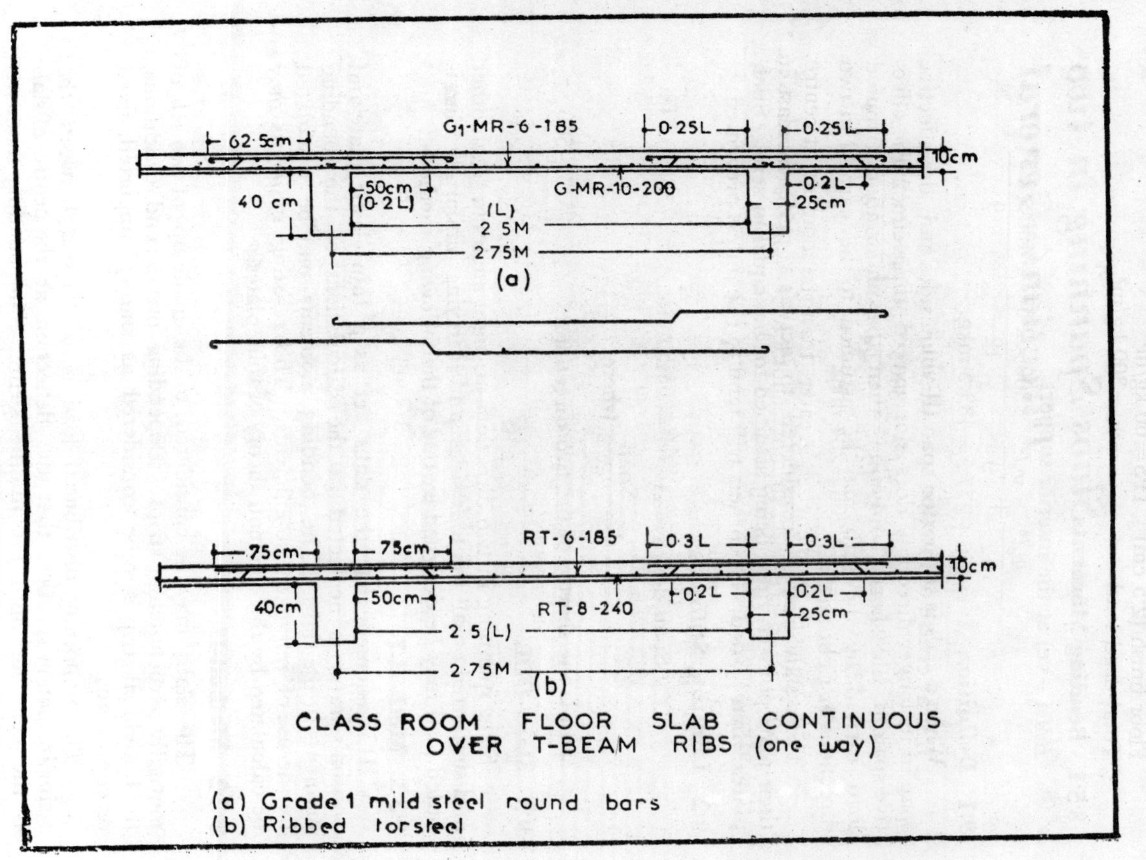

CLASS ROOM FLOOR SLAB CONTINUOUS OVER T-BEAM RIBS (one way)

(a) Grade 1 mild steel round bars
(b) Ribbed torsteel

9

Slabs Spanning in two Directions–General

9·1 Definition

When a slab is supported on all four sides and the effective span in its longer direction does not exceed three times the effective span in the shorter direction, then the uniformly distributed load carried by the slab may be assumed to be carried in two directions at right angles.

These slabs may be considered to act as a perfectly elastic, their plate poissons ratio being assumed to be equal to zero. Such plates when loaded, the corners will tend to rise unless prevented.

9·2 Limiting Stiffness

Span/depth=35

i.e., $d=\frac{\text{Span}}{35}$ where

span is the shorter of the two directions.

9·3 Slabs UDL

Slabs spanning in two directions at right angles to each other and supported on all the four sides and carrying uniformly distributed load may be designed by one of the following methods :

9·3·1 Method-I

It is assumed that the slabs act as a thin elastic plate and poisson's ratio is neglected in the calculations of the bending moments in the slabs. The bending moments may be calculated by the use of influence co-efficients. The resisting moments may be calculated by the common theory of thin plates.

9·3·2 Method-II

The slabs may be considered to be made up of two sets of mutually perpendicular strips. Depending on the end conditions, the individual strip is to be considered as simply supported, fixed or continuous.

The uniformly distributed load w is carried between the strips in such proportions that the deflection at the centre of slab is the same for both the middle strips. The bending moments calculated on this basis have to be corrected for torsion and corner restraint.

Mid span moments shall be calculated using the following expressions :

BENDING MOMENT COEFFICIENTS FOR RECTANGULAR PANELS SUPPORTED ON FOUR SIDES WITH PROVISION FOR TORSION AT CORNERS

Type of panel and moments	Bending moment Z'_x for short span for values of $\frac{l_y}{l_x}$ =								B.M. coefficient Z'_y for long span for all values of $\frac{l_y}{l_x}$.
	1.0	1.1	1.2	1.3	1.4	1.5	1.75	2.0	
i) Interior panels									
Neg. moment at continuous edge	0.033	0.040	0.045	0.050	0.054	0.659	0.011	0.083	0.033
ii) Positive moment at mid span	0.25	0.030	0.034	0.038	0.041	0.045	0.053	0.062	0.025
iii) One short or long edge discontinuous									
Neg. moment at continuous edge	0.041	0.047	0.053	0.057	0.061	0.065	0.075	0.085	0.041
iv) Positive moment at mid span	0.031	0.035	0.040	0.043	0.046	0.049	0.056	0.064	0.031
iii) Two adjacent edges discontinuous									
Neg. moment at continuous edge	0.049	0.056	0 062	0.066	0.070	0.073	0.082	0.090	0 049
Positive moment at mid span	0.037	0.042	0.047	0.050	0.053	0.055	0.062	0.068	0.037

iv) Two short edges discontinuous									
Neg. moment at continuous edge	0.056	0.061	0.065	0.069	0.071	0.073	0.077	0.080	—
Positive moment at mid span	0.044	0.046	0.049	0.051	0.053	0.055	0.058	0.060	0.044
v) Two long edges discontinuous									
Neg. moment at continuous edge	—	—	—	—	—	—	—	—	—
Positive moment at mid span	0.044	0.053	0.060	0 065	0.068	0.071	0 077	0.080	0.044
vi) Three edges discontinuous (one short or long edge continuous)									
Neg. moment at continuous edge	0.058	0.065	0.071	0.077	0.081	0.085	0.092	0.099	0.058
Positive moment at mid span	0.044	0.049	0.054	0.058	0.061	0.064	0 069	0.074	0.044
vii) Four edges discontinuous									
Positive moment at mid span	0.050	0.057	0.062	0.067	0.071	0.075	0.081	0.083	0.050

Coefficients for two-way reinforced slab with and without Torsion reinforcement

Moment $\frac{l_y}{l_x}$	1 l_x	m_x	m_y	2 l_{x_2}	m_{x_2}	m_{y_3}	3 l_{x_3}	m_{x_3}	m_{y_3}	4 l_{x_4}	m_{x_4}	m_{y_4}	5 l_{x_5}	m_{x_5}	m_{y_5}	6 l_{x_6}	m_{x_6}	m_{y_6}
0.6	0.115	94.9 80.5	12.3 10.4	0.245	85.2 69	13.7 11.9	0.393	87.6	16.1	0.115	146 133	18.9 17.4	0.206	138	20.7	0.115	229	29.8
0.7	0.194	61.6 49.5	14.8 11.9	0.375	59.1 46.3	17.2 14.7	0.546	63.7	21.6	0.194	90.2 80.7	21.6 19.5	0.324	90.7	74.9	0.194	139	33.4
0.8	0.291	40.3 33.9	18.1 13.9	0.506	44.7 34.6	21.9 18.7	0.672	50.4	29.6	0.291	62.2 54.6	25.4 22.4	0.450	66.3	31.0	0.291	94.5	38.8
0.9	0.396	34.1 25.4	22.4 16.6	0.621	35.7 27.9	28.4 24.2	0.766	42.5	40.6	0.396	46.6 40.5	30.6 26.6	0.568	52.5	39.4	0.396	70.1	46.0
1.0	0.500	27.4 20.2	27.4 20.2	0.714	29.9 23.9	36.08 31.8	0.833	37.5	57.7	0.500	37.2 32.2	37.2 32.2	0.667	42.2	50.6	0.500	57.7	55.7
1.1	0.594	22.8 16.9	33.3 24.8	0.785	26.0 21.4	47.5 42.0	0.880	34.2	75.5	0.594	31.1 27.0	45.5 27.6	0.745	38.9	65.2	0.594	46.8	68.5

1.2	0.675	19.4 / 14.7	40.3 / 30.6	0.838	23.3 / 19.6	61.5 / 55.0	0.912	31.9	102	0.675	27.0 / 23.7	56.0 / 49.2	0.806	35.3	84.2	0.675	40.9	85.5
1.3	0.741	17.0 / 13.2	48.5 / 37.6	0.877	21.4 / 18.4	78.8 / 71.0	0.934	30.3	135	0.741	74.2 / 21.4	69.0 / 61.0	0.51	32.8	108	0.741	36.9	105
1.4	0.794	15.2 / 12.1	58.4 / 46.6	0.906	20.0 / 17.6	100.0 / 91.5	0.951	29.2	176	0.794	22.1 / 38.1	85.0 / 76.6	0.885	31.0	138	0.794	34.1	131
1.5	0.835	13.9 / 11.3	70.2 / 57.5	0.927	19.0 / 17.0	127.0 / 118	0.961	28.3	227	0.835	20.6 / 18.7	104.0 / 95.0	0.910	29.7	175	0.835	32.0	162
1.6	0.868	12.9 / 10.8	84.5 / 70.6	0.942	18.2 / 16.5	159 / 148	0.970	27.6	288	0.868	19.5 / 17.8	128 / 117	0.929	28.7	219	0.868	30.6	200

Note :—L_x=load coefficient for edge moment in l_x direction.

$(1\text{-}l_x)$=load coefficient for edge moment in l_y direction.

Bottom figures for cases 1, 2 and 4 are the values of m_x and m_y when torsion reinforcement at the corner of discontinuous edges is not provided.

$$M_x = \frac{wl^2_x}{m_x}$$

$$M_y = \frac{wl^2_y}{m_y}$$

where l_x, l_y=length of the sides of the slab

m_x, m_y=Co-efficients in the Table 9·3·2.

If the slab is fixed at edge beam or is continuous no torsional reinforcement need be provided at the corners. In the case of those slabs which are neither fixed at some of its edges nor continuous, the torsional reinforcement need not be provided, if the slab is designed for increased bending moments as in Table 9·3·2.

Slabs in which the edges are discontinuous, top and bottom reinforcement shall be provided for a length equal to one-fifth of the span in each direction. Both top and bottom reinforcement shall consist of two layers of bars placed either perpendicular or parallel to the respective slab edges. The areas of reinforcement in each of the four layers, per unit width of the slab shall be equal to three quarters of the main reinforcement at mid span.

9·3·3 Method-III

This method is based on mathematical analysis, supplemented by experimental data. The effects of redistribution of bending moments have been partially allowed for and are computed from the expressions

$$M_x = Z'_x \, wl^2_x$$
$$M_y = Z'_y \, wl^2_y$$

where Z'_x, Z'_y are the co-efficients from the Table 9·3·3 for different ratios of $\frac{l_y}{l_x}$.

9·4 Slabs carrying concentrated Loads

For determining bending moments in slabs spanning in two directions at right angles and carrying concentrated load, any accepted method approved by the engineer in-charge may be adopted.

10

Simply Supported Slab Spanning in two Directions

10·1 Data

Drawing room in a residence
size=4·3 m × 5·6 m
Edge conditions—simply supported on all four edges on load bearing walls of 39 cm thick and discontinuous.
Materials : Ribbed torsteel, Concrete M 150
Design the slab by, (*i*) Rankine Grashoff theory
(*ii*) ISI code method.

10·2 Relevant Codes

ISI—456, ISI—875

10·2·1 Allowable Stresses

Using Mild steel

$C=50$ kg/cm², $q_b=10$ kg/cm², $q_s=5$ kg/cm²

$t=1400$ kg/cm², $m=19$, $R=8{\cdot}74$ $a={\cdot}865\ d_e$

10·3 Type

$l_x=4{\cdot}3$ m, $l_y=5{\cdot}6$ m

$$l_y/l_x=\frac{5{\cdot}6}{4{\cdot}3}=1{\cdot}3$$

Slab spans in two directions and is simply supported on all edges.

10·4 Slab Depth

For slabs simply supported spanning in two directions

$$d=\frac{\text{span}}{35}=\frac{430}{35}=12{\cdot}3 \text{ cm}$$

Assume 15 cm.

10·4·1 Effective Depth

$d_e=15-$(clear cover$+\frac{1}{2}$ the dia of bar)

Using 12 mm ϕ bars

$d_e=13{\cdot}1$ cm Say 13 cm

10·4·2 Effective Span

l_{xe}=c/c to supports=4·3+0·3=4·6 m

=clear span+d_e=4·3+0·13=4·43 m

l_{ye}=5·6+·3=5·9 m

or =5·6+0·12 (assuming 10 mm ϕ)=5·72 m

10·5 Loads

Live load on roof (Residence)=150 kg/m²
Self weight of slab (15 cm) =360 kg/m²
Brick bat concrete roof finish =100 kg/m²
Total load=w=610 kg/m²

10·5·1 Load Distribution—Rankine—Grashoff method

Load carried by 1 m strip along the shorter span

$$w_x=\frac{wl^4_y}{l^4_x+l^4_y}=\frac{610\times 5{\cdot}72^4}{4{\cdot}43^4+5{\cdot}72^4}=446 \text{ kg/m}$$

Load carried by longer span

$$w_y=610-446=164 \text{ kg/m}$$

10·5·2 Bending Moments

$$M_x=\frac{w_x\times l^2_x}{8}=\frac{446\times 4{\cdot}43^2}{8}=1090 \text{ mkg}$$

$$M_y=\frac{w_y l^2_y}{8}=\frac{164\times 5{\cdot}72^2}{8}=675 \text{ mkg}$$

10·5·3 Effective Depth

$$d_e=\sqrt{\frac{1090\times 100}{100\times 8{\cdot}74}}=\sqrt{124{\cdot}2}=11{\cdot}1 \text{ cm}$$

Use $d=13{\cdot}5$ cm

$$d_e=13{\cdot}5-(1{\cdot}3+1/2\times 1{\cdot}2)=11{\cdot}6 \text{ cm}$$

10·5·4 Main Steel

Steel for shorter span

$$A_t=\frac{1090\times 100}{1400\times{\cdot}865\times 11{\cdot}6}=7{\cdot}8 \text{ cm}^2$$

Use 12 mm at 14·5 cm c/c

Steel for longer span

Assuming 10 mm ϕ bars d_e=10·5 cm

$$A_t=\frac{67500}{1400\times{\cdot}865\times 10{\cdot}5}=5{\cdot}3 \text{ cm}^2$$

Use 10 mm ϕ at 14·5 cm c/c

10·5·5 Effective Depth using torsteel

$$d_e=\sqrt{\frac{1090\times 100}{100\times 6{\cdot}59}}=\sqrt{166}=12{\cdot}9 \text{ cm}$$

$$d=15 \text{ cm},\quad d_e=13 \text{ cm}$$

10·5·6 Main Steel

$$A_t \text{ for short span}=\frac{109000}{2300\times{\cdot}903\times 13}=4{\cdot}05 \text{ cm}^2$$

Use 10 mm ϕ at 19 cm c/c

Using 8 mm, $d_e=12{\cdot}3$ cm

$$A_t \text{ for longer span}=\frac{67500}{2300\times{\cdot}903\times 12{\cdot}3}=2{\cdot}62 \text{ cm}^2$$

Use 8 mm ϕ at 19 cm c/c

10·6 ISI Code Method-I

$$l_x=4{\cdot}43 \text{ m},\quad l_y=5{\cdot}72 \text{ m}$$

$$\frac{l_y}{l_x}=\frac{5{\cdot}72}{4{\cdot}43}=1{\cdot}3$$

Slab is simply supported on all four edges and also discontinuous

10·6·1 B.M. Coefficients

For the ratio $\frac{l_y}{l_x}=1{\cdot}3$

Z'_x for shorter span$=0{\cdot}067$

Z'_y for longer span$={\cdot}05$

10·6·2 Bending Moments

$$M_x=Z'_x wl^2_x={\cdot}067\times 610\times 4{\cdot}43^2=801 \text{ mkg}$$

$$M_y=Z'_y wl^2_x={\cdot}05\times 610\times 4{\cdot}43^2=600 \text{ mkg}$$

10·6·3 Effective Depth

Using grade—1 steel

$$d_e=\sqrt{\frac{801\times 100}{100\times 8{\cdot}74}}=\sqrt{92}=8{\cdot}82 \text{ cm}$$

Use $d_e=9$ cm, $d=11$ cm

10·6·4 Main Steel

$$A_t \text{ for shorter span}=\frac{80100}{1400\times{\cdot}865\times 9}=7{\cdot}35 \text{ cm}^2$$

Use 12 mm at 15 cm c/c

Using 10 mm ϕ bars, $d_e=8$ cm

A_t for longer span $= \dfrac{60000}{1400 \times \cdot 865 \times 8} = 6 \cdot 22 \text{ cm}^2$

Use 10 mm ϕ at 12·5 cm c/c

10·6·5 Effective Depth using ribbed torsteel

$$d_e = \sqrt{\frac{80100}{100 \times 6 \cdot 59}} = \sqrt{122} = 11 \cdot 01$$

$$d = 13 \text{ cm}, \qquad d_e = 11 \text{ cm}$$

10·6·6 Main Steel

A_t for short span $= \dfrac{80100}{2300 \times \cdot 903 \times 11} = 3 \cdot 5 \text{ cm}^2$

Use 10 mm ϕ at 22 cm c/c

A_t for longer span $= \dfrac{60000}{2300 \times \cdot 903 \times 10 \cdot 2} = 2 \cdot 82 \text{ cm}^2$

Use 10 mm ϕ at 27 cm c/c

10·7 ISI Method II Coefficients

Grade—I steel

$m_x = 13 \cdot 2$, $m_y = 37 \cdot 6$, allowing for torsion in the corners

$w = 613 \text{ kg/m}^2$

10·7·1 Bending Moments

$$M_x = \frac{wl^2_x}{m_x} = \frac{610 \times 4 \cdot 43^2}{13 \cdot 2} = 910 \text{ mkg}$$

$$M_y = \frac{wl^2_y}{m_y} = \frac{610 \times 5 \cdot 72^2}{37 \cdot 6} = 532 \text{ mkg}$$

10·7·2 Effective Depth

$$d_e = \sqrt{\frac{910 \times 100}{100 \times 8 \cdot 74}} = \sqrt{103} = 10 \cdot 15 \text{ cm}$$

Use $\quad d = 12 \cdot 5 \text{ cm} \qquad d_e = 10 \cdot 5 \text{ cm}$

10.7.3 Main Steel

A_t for shorter span $- \dfrac{91000}{1400 \times \cdot 865 \times 10 \cdot 5} - 7 \cdot 15 \text{ cm}^2$

Use 12 mm ϕ at 15·5 cm c/c

A_t for longer span $= \dfrac{53200}{1400 \times \cdot 865 \times 9 \cdot 5} = 4 \cdot 65 \text{ cm}^2$

Use 10 mm ϕ at 16·5 cm c/c

10·7·4 Effective Depth using ribbed steel

$$d_e = \sqrt{\frac{91000}{100 \times 6{\cdot}59}} = \sqrt{138} = 11{\cdot}8 \text{ cm}$$

Use $d = 14$ cm, $d_e = 12$ cm

A_t for shorter span $= \dfrac{91000}{2300 \times {\cdot}903 \times 12} = 3{\cdot}65 \text{ cm}^2$

Use 10 mm ϕ at 21 cm c/c

A_t for longer span $= \dfrac{53200}{2300 \times {\cdot}903 \times 11{\cdot}2} = 2{\cdot}28 \text{ cm}^3$

Use 8 mm ϕ at 22 cm c/c

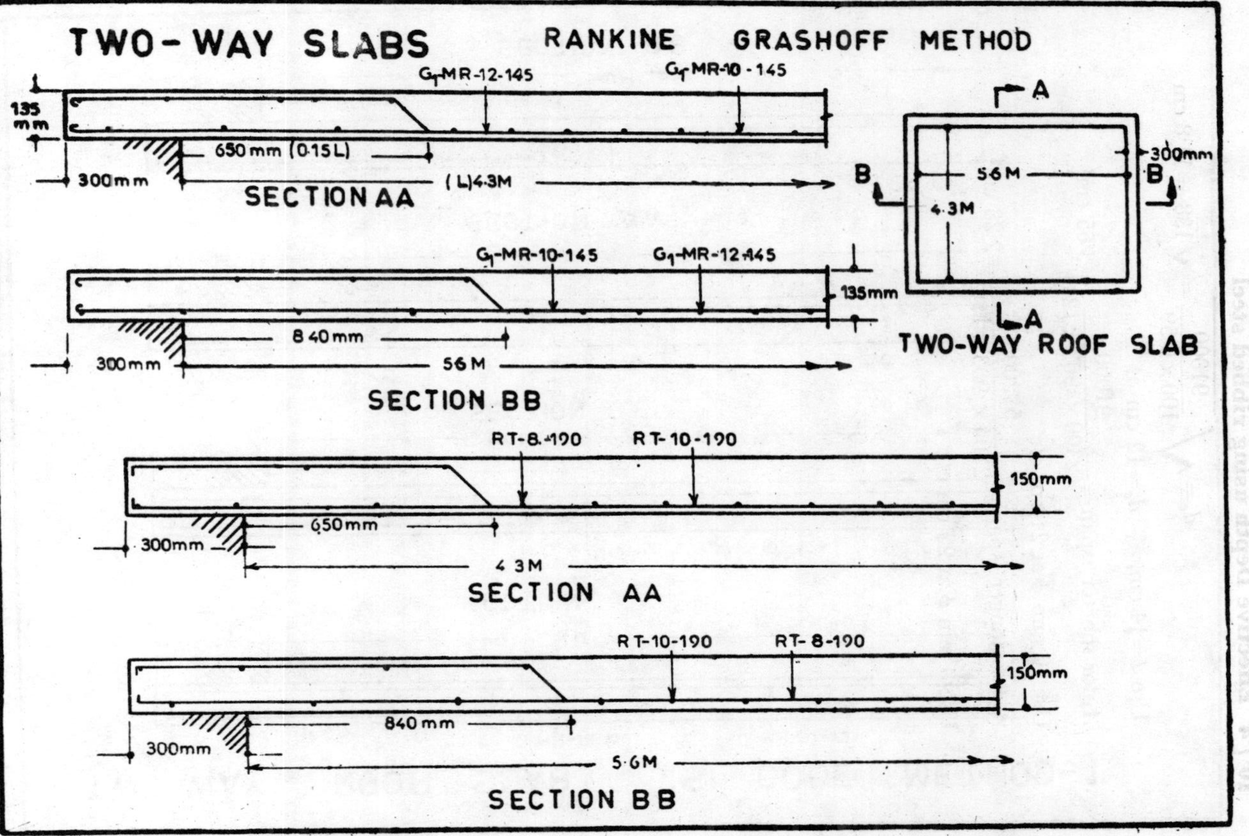

TWO - WAY SLABS
RANKINE GRASHOFF METHOD
G1-MR-12-145
G1-MR-10-145
135 mm
650 mm (0.15L)
300mm
(L)4.3M
SECTION AA
G1-MR-10-145
G1-MR-12-145
135mm
8 40mm
300mm
56M
SECTION BB
A
B
300mm
56M
4.3M
TWO-WAY ROOF SLAB
RT-8-190
RT-10-190
150mm
650mm
300mm
4 3M
SECTION AA
RT-10-190
RT-8-190
150mm
840mm
300mm
5.6M
SECTION BB

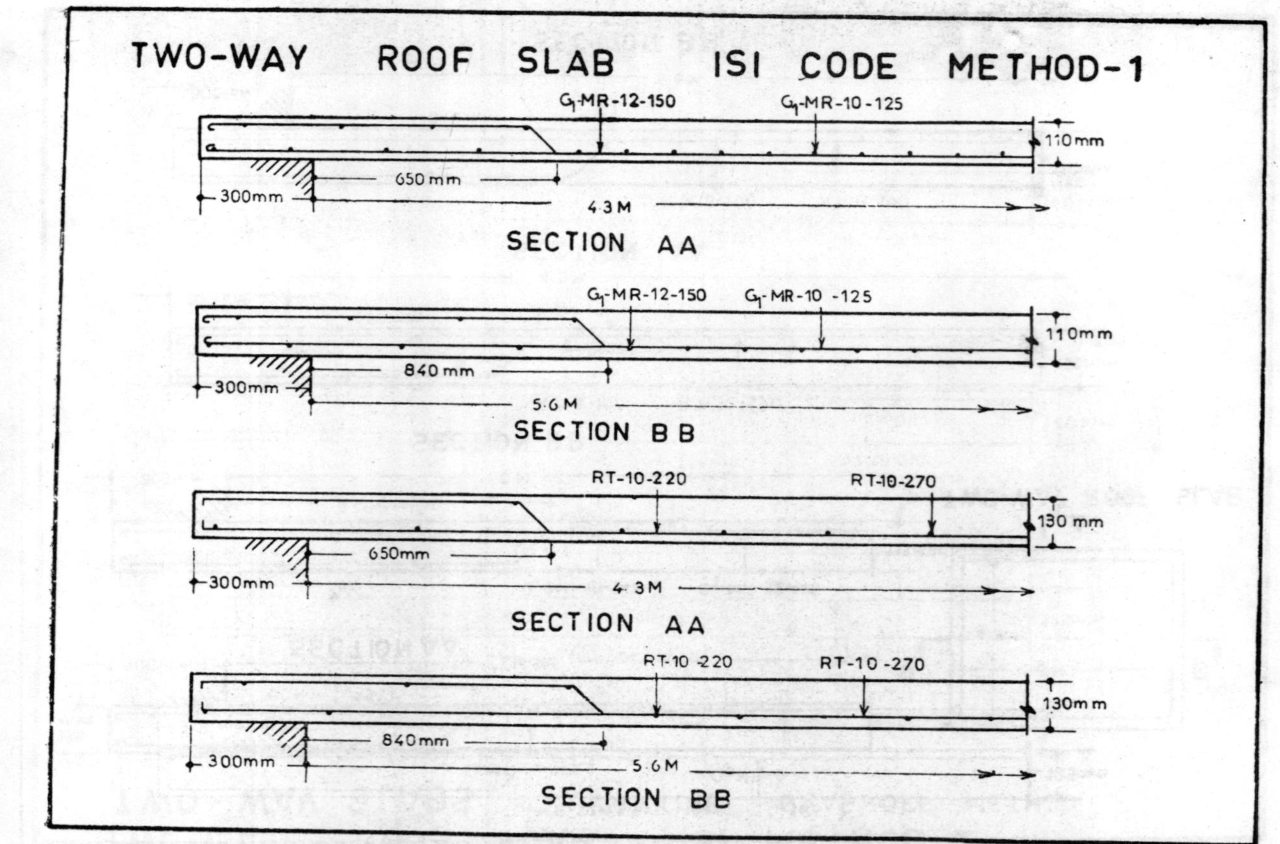
TWO-WAY ROOF SLAB ISI CODE METHOD-1
G1-MR-12-150
G1-MR-10-125
110mm
650 mm
300mm
4.3 M
SECTION AA
G1-MR-12-150
G1-MR-10-125
110mm
840 mm
300mm
5.6 M
SECTION BB
RT-10-220
RT-10-270
130 mm
650mm
300mm
4.3 M
SECTION AA
RT-10-220
RT-10-270
130mm
840mm
300mm
5.6 M
SECTION BB

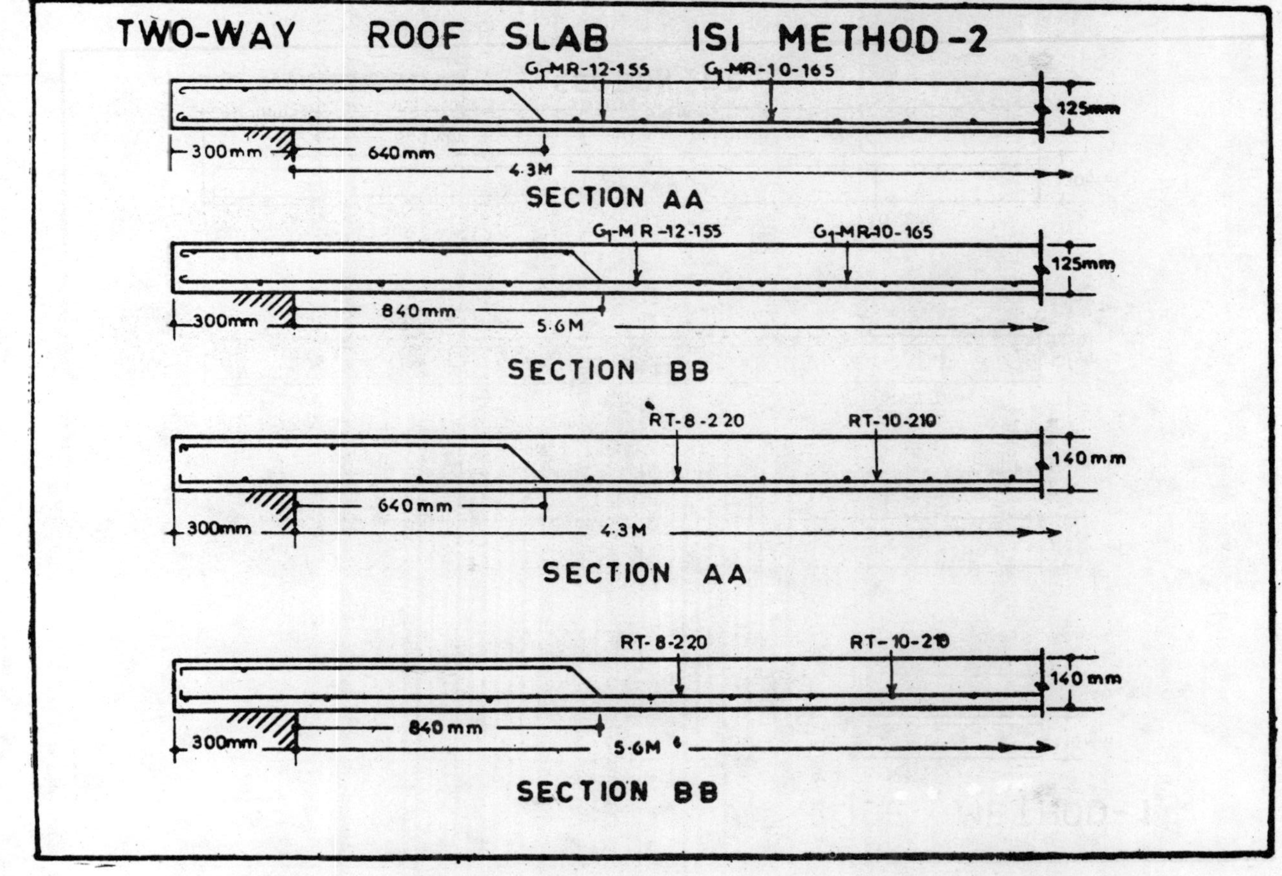

TWO-WAY ROOF SLAB ISI METHOD-2
G-MR-12-155
G-MR-10-165
125mm
300mm
640mm
4·3M
SECTION AA
G-MR-12-155
G-MR-10-165
125mm
300mm
840mm
5·6M
SECTION BB
RT-8-220
RT-10-210
140mm
300mm
640mm
4·3M
SECTION AA
RT-8-220
RT-10-210
140mm
300mm
840mm
5·6M
SECTION BB

11

Cantilever Slab–General

11·1 Definition

Slabs in which only one edge is restrained and the other three are free, are called cantilever slabs.

11·2 Limiting stiffness

Span/depth = 12

11·3 Effective span

The effective over hang should be taken as the effective span.

In the design of cantilever slabs the specifications for slabs spanning in one direction are applicable.

12

Chajja at Window Sill Level Carrying Louvres

12·1 Data

Span=length of chajja projection=0·9 m

Louvre Size=50 cm×1·6 m×5 cm

Number of Louvres supported on chajja=4

Width of chajja=1·3 m

Materials : M 150 concrete, Grade—1 Steel

12·2 Relevant Codes

IS—456, IS—875

12·2·1 Allowable Stresses

$C=50$ kg/cm², q_b average$=6$ kg/cm², $t=1400$ kg/cm²
$m=19$, $R=8{\cdot}74$, $a=0{\cdot}865\ d_e$

12·3 Depth of Slab

$$d=\frac{l}{12}=\frac{90}{12}=7{\cdot}5 \text{ cm}$$

Assume $d=10$ cm, $d_e=7{\cdot}5$ cm

12·3·1 Effective Span

$$l=0{\cdot}9 \text{ m}$$

12·4 Loads

$$\text{Dead load due to slab}=\frac{10\times2400\times1}{100}=240 \text{ kg/m}^2$$

$$\text{Dead load due to louvres}=\frac{5\times1{\cdot}6\times50\times4\times2400}{100\times100\times1{\cdot}3\times0{\cdot}9}=326 \text{ kg/m}^2$$

Total load $w=566$ kg/m²

12·4·1 Bending Moment

$$M=\frac{wl^2}{2}=\frac{566\times{\cdot}9\times{\cdot}9}{2}=230 \text{ mkg.}$$

12·4·2 Effective Depth

$$d_e=\sqrt{\frac{23000}{100\times8{\cdot}74}}=\sqrt{26{\cdot}4}=5{\cdot}15 \text{ cm}$$

Use $d=10$ cm and $d_e=8$ cm

12·4·3 Main Steel

$$A_t = \frac{23000}{1400 \times \cdot 865 \times 8} = 2{\cdot}38 \text{ cm}^2$$

Use 6 mm ϕ at 11·5 cm c/c

12·4·4 Secondary Steel

$$A_s = \frac{\cdot 15}{100} \times 100 \times 10 = 1{\cdot}5 \text{ cm}^2$$

Use 6 mm ϕ at 18·5 cm c/c

12·4·5 Anchorage Length

Anchorage length $= n \times$ diameter of bar $= \frac{t}{4 \times q_b} \times$ dia. of bar

$$= \frac{1400 \times 0{\cdot}6}{4 \times 6} = 58{\cdot}3 \times 0{\cdot}6 = 35 \text{ cm}$$

Provided length = 60 cm.

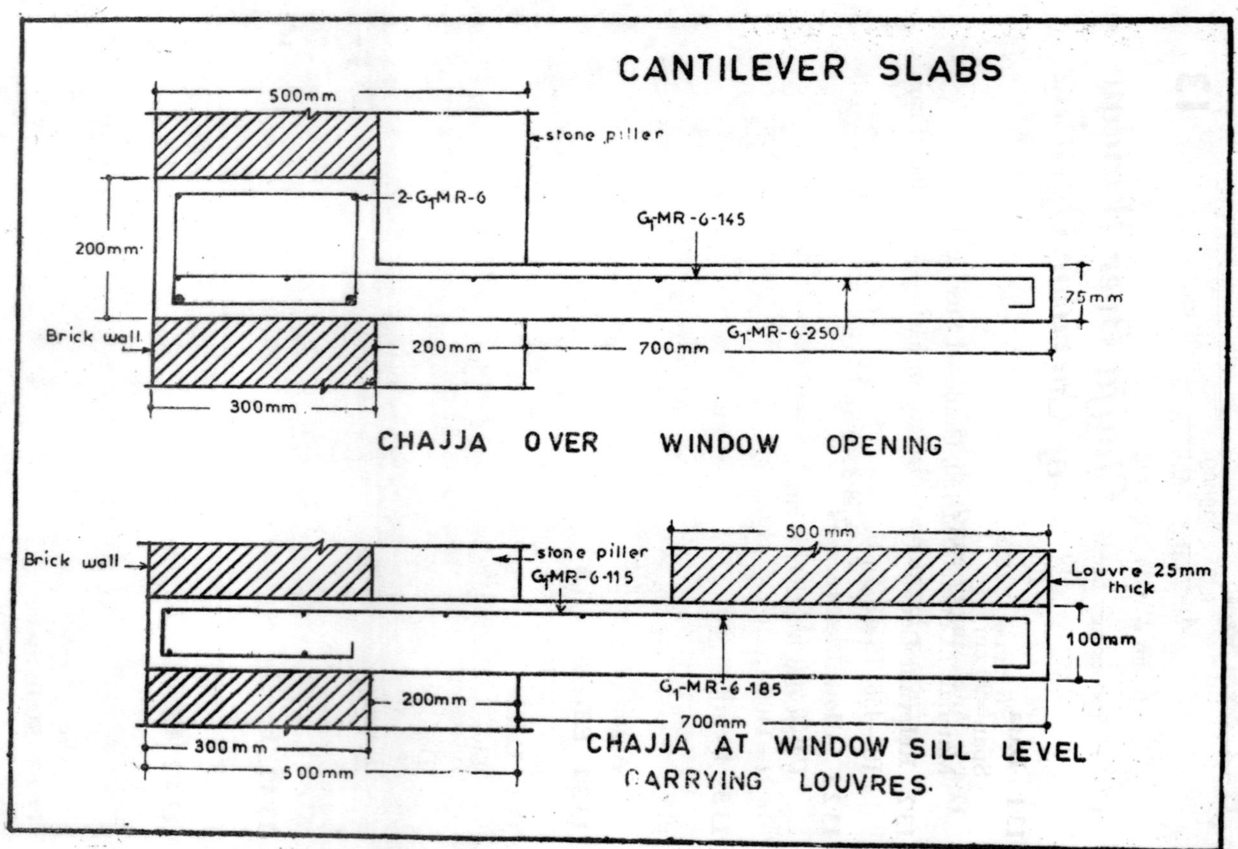
CANTILEVER SLABS
500mm
stone piller
2-G1-MR-6
G1-MR-6-145
200mm
75mm
Brick wall
G1-MR-6-250
200mm
700mm
300mm
CHAJJA OVER WINDOW OPENING
500 mm
Brick wall
stone piller
G1-MR-6-115
Louvre 25mm thick
100mm
G1-MR-6-185
200mm
700mm
300mm
500mm
CHAJJA AT WINDOW SILL LEVEL
CARRYING LOUVRES.

13

Chajja Over Window of Corridor Opening

13·1 Data

Span=90 cm

Materials : Concrete M 150, Grade—1 Steel

13·2 Relevant Codes

IS—456, IS—875

13·2·1 Allowable Stresses

$C=50$ kg/cm², $q_s=5$ kg/cm², $q_b=10$ kg/cm²

$t=1400$ kg/cm², $m=19$, R=8·74, $a=0{\cdot}865\ d_e$

13·3 Depth of Slab

$$d=\frac{l}{12}=\frac{90}{12}=7{\cdot}5 \text{ cm}$$

Assume $d=7{\cdot}5$ cm, $d_e=5{\cdot}5$ cm.

13·3·1 Effective Span

$l=0{\cdot}9$ m

13·4 Loads

Live load (inaccessible roof)=75 kg/cm²

Dead load (7·5 cm)$=\frac{7{\cdot}5}{100}\times1\times2400=180$ kg/m²

Plastering (2·5 cm)=52 kg/cm²

Total load=307 kg/m² Say=310 kg/cm²

$w=310$ kg/m²

13·4·1 Bending Moments

$$M=\frac{wl^2}{2}=\frac{310\times0{\cdot}9\times0{\cdot}9}{2}=126 \text{ mkg}$$

13·4·2 Effective Depth

$$d_e=\sqrt{\frac{12600}{100\times8{\cdot}74}}=\sqrt{14{\cdot}4}=3{\cdot}8 \text{ cm}$$

Use $d=7{\cdot}5$ cm, $d_e=5{\cdot}5$ cm

13·4·3 Main Steel

$$A_t=\frac{12600}{1400\times0{\cdot}865\times5{\cdot}5}=1{\cdot}9 \text{ cm}^2$$

Use 6 mm ϕ bar at 14·5 cm c/c

13·4·4 Secondary Steel

$$A_s=\frac{0\cdot15}{100}\times7\cdot5\times100=1\cdot125\ \text{cm}^2$$

Use 6 mm ϕ bar at 25 cm c/c

13·4·5 Anchorage Length

Anchorage length $= n\times$ diameter of bar

$$n=\frac{t}{4q_b}=\frac{1400}{4\times6}=58\cdot33$$

Anchorage length $=58\cdot33\times0\cdot6=35$ cm

14

Circular Slabs–General

14·1 Simply Supported

Radial moments

M_{r_c} = Maximum radial moment at centre

$$= \frac{3}{16} wR^2$$

M_{r_s} = Maximum radial moment at support = 0

M_{c_s} = Maximum circumferential moment at support

$$= \frac{2}{16} wR^2$$

14·2 Fixed

Radial moments

M_{r_c} = Maximum radial moment at centre (Positive)

$$= \frac{wR^2}{16}$$

M_{r_s} = Maximum radial moment at support (Negative)

$$= \frac{2}{16} wR^2$$

M_{c_c} = Maximum circumferential moment at centre

$$= \frac{wR^2}{16}$$

M_{c_s} = Maximum circumferential moment at support = 0

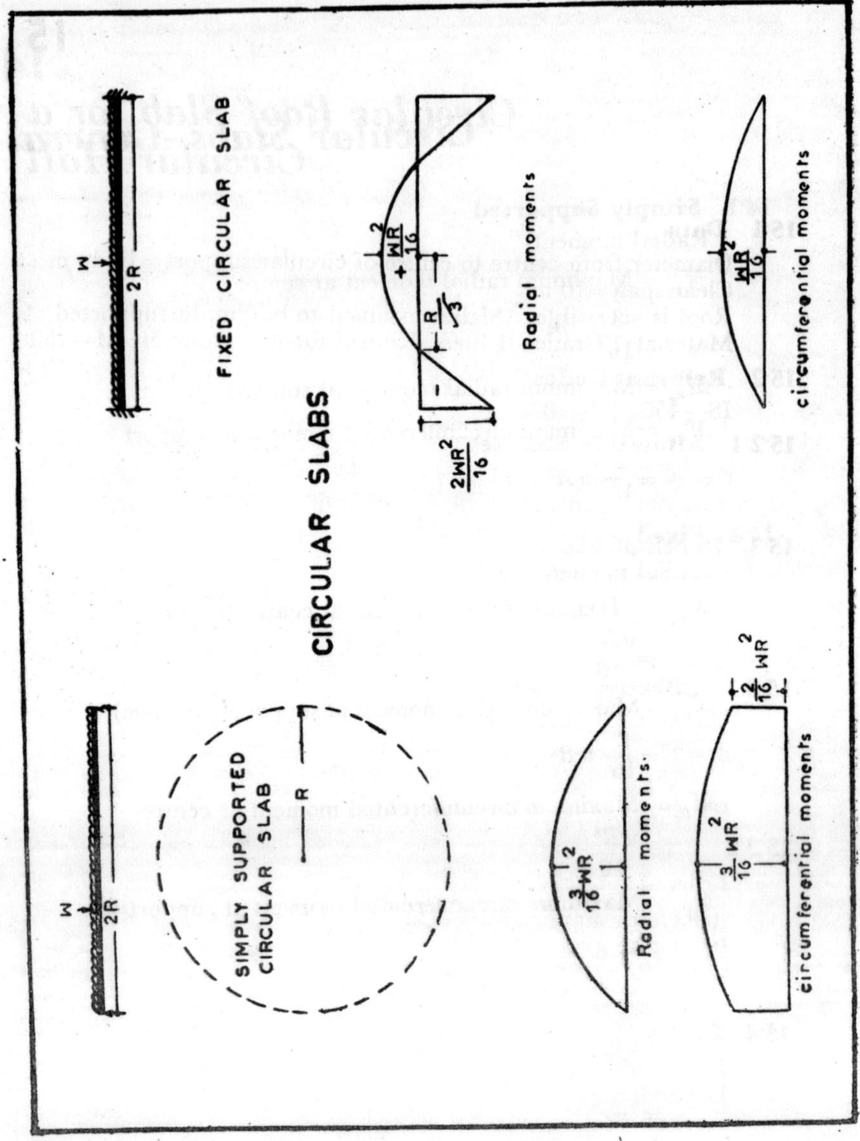

CIRCULAR SLABS
SIMPLY SUPPORTED CIRCULAR SLAB
W
2R
R
Radial moments
$\frac{3}{16}WR^2$
circumferential moments
$\frac{3}{16}WR^2$
$\frac{2}{16}WR^2$
FIXED CIRCULAR SLAB
W
2R
Radial moments
$+\frac{WR^2}{16}$
$\frac{R}{\sqrt{3}}$
−
$\frac{2WR^2}{16}$
circumferential moments
$\frac{WR^2}{16}$

15

Circular Roof Slab for a Circular Hall

15·1 Data

Diameter from centre to centre of circular support = 10·40 m
Clear span = 10 m
Roof is accessible. Slab is assumed to be simply supported.
Materials : Grade—1 Steel, Ribbed torsteel, Concrete M—150

15·2 Relevant Codes

IS—456. IS—875

15·2·1 Allowable Stresses

$C=50$ kg/cm², $q_s=5$ kg/cm², $q_b=10$ kg/cm²
$t=1400$ kg/cm² $R=8{\cdot}74$, $a=0{\cdot}865\ d_e$, $m=19$

15·3 Depth of Slab

For circular slab

$$d=\frac{1}{40}\text{ of diameter}=\frac{1}{40}\times 1000=25\text{ cm}$$

15·3·1 Effective Depth

Using 16 mm dia bars

$d_e=25-(\text{Cover}+\frac{1}{2}\text{ the diameter of the bar})$
$=25-(1{\cdot}3+\frac{1}{2}\times 1{\cdot}6)=25-2{\cdot}1=22{\cdot}9$ cm
try $d=25$ cm, $d_e=22{\cdot}5$ cm

15·3·2 Effective Diameter

Least of

(*i*) c/c of supports = 10·40 m
(*ii*) Clear span + eff : depth = 10 + ·225 = 10·225 m
$2R=10{\cdot}225$ m
$R=5{\cdot}1125$ m Say = 5·113 m

15·4 Loads

Live load on accessible roofs = 150 kg/m²
Dead load of slab (25 cm) = 25 × 24 = 600 kg/m²
Roof finish with coba for water proofing = 150 kg/m²
Total load $w=900$ kg/m²

15·4·1 Bending Moments

Radial

$$M_{r_c}=\frac{3}{16}\ wR^2=3/16\times 900\times 5{\cdot}113^2=4440\text{ mkg}$$

15·4·2 Circumferential

$$M_{C_c} = \frac{3}{16}\, wR^2 = 4440 \text{ mkg}$$

$$M_{C_s} = \frac{2}{16}\, wR^2 = 2960 \text{ mkg}$$

15·4·3 Effective Depth

$$d_e = \sqrt{\frac{4440 \times 100}{100 \times 8{\cdot}74}} = \sqrt{508} = 22{\cdot}5 \text{ cm}$$

Use $d = 25$ cm $d_e = 22{\cdot}5$ cm

15·4·4 Main Steel for Radial Moment

$$A_t = \frac{4440 \times 100}{1400 \times .865 \times 22{\cdot}5} = 16{\cdot}3 \text{ cm}^2$$

Use 16 mm ϕ bars at 12 cm c/c in both the directions.

15·4·5 Secondary Steel

Main steel used in either direction is greater than 0·15% of concrete section. Therefore there is no need for secondary steel.

15·4·6 Main Steel for Circumferential Moment

$$A_t = \frac{2960 \times 100}{1400 \times .865 \times 22{\cdot}5} = 10{\cdot}9 \text{ cm}^2$$

Use 16 mm ϕ bar at 18 cm c/c at the edge of the slab for a length of $R/4 = \frac{5{\cdot}113}{4} = 1{\cdot}28$ m in the form of circular rings.

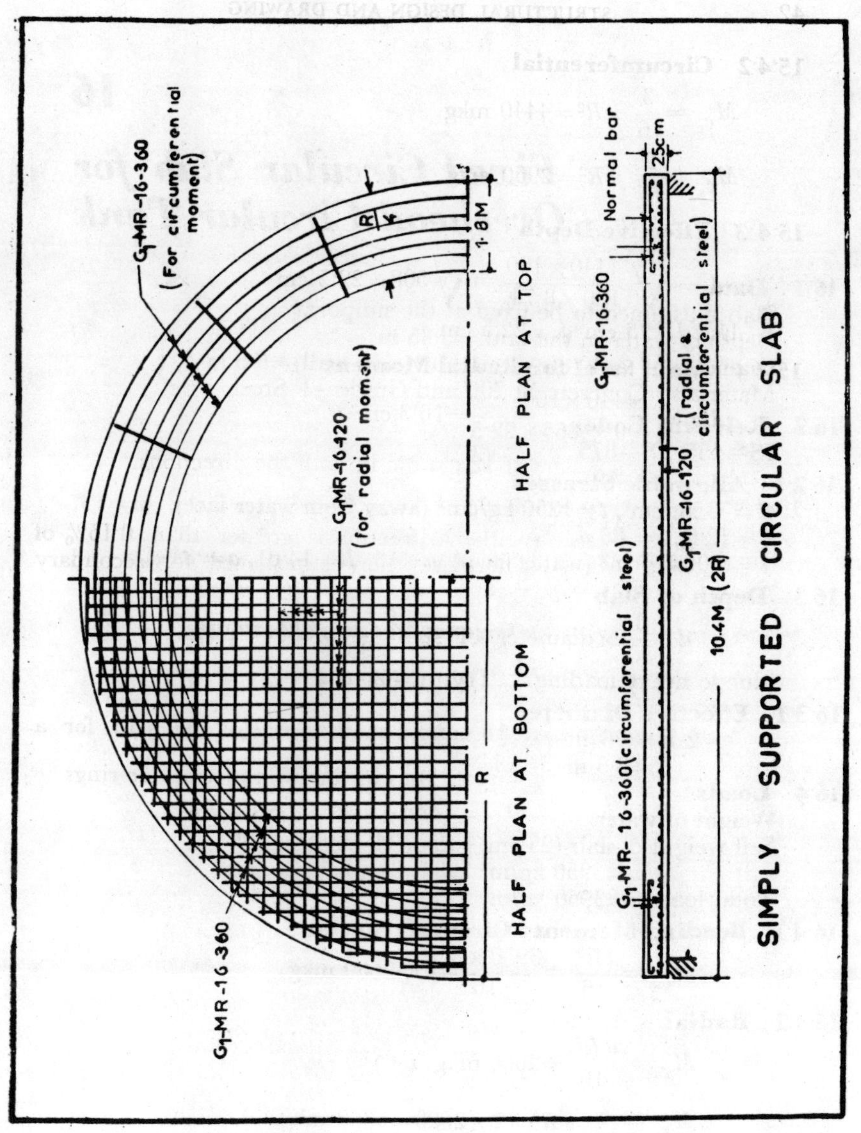
G1-MR-16-360
(For circumferential moment)
R
1·8 M
G1-MR-16-120
(for radial moment)
HALF PLAN AT TOP
HALF PLAN AT BOTTOM
R
G1-MR-16 360
Normal bar
25cm
G1-MR-16-360
G1-MR-16-120
(radial & circumferential steel)
G1-MR- 16-360(circumferential steel)
10.4M (2R)
SIMPLY SUPPORTED CIRCULAR SLAB

16

Fixed Circular Slab for Overhead Circular Tank

16·1 Data

Slab is assumed to be fixed at the support.
Depth of water in the tank=3·35 m
Diameter of tank (c/c of circular side wall)=6·6 m
Materials : Concrete M 200 and Grade—1 Steel

16.2 Relevant Codes

IS—456, IS—875

16·2·1 Allowable Stresses

$C=70$ kg/cm², $t=1250$ kg/cm² (away from water face), $m=13$, $R=12{\cdot}6$, $a=.86\ d_e$

$t=1000$ kg/cm² (water face), $m=13$, $R=14{\cdot}01$, $a={\cdot}84\ d_e$.

16·3 Depth of Slab

$$d=\frac{1}{30}\times\text{diameter of slab}=\frac{1}{30}\times660=22\text{ cm.}$$

(due to heavy loading). Try $d=25$ cm

16·3·1 Effective Diameter

$2R=6{\cdot}6$ m

$R=3{\cdot}3$ m

16·4 Loads

Weight of water $=1000\times3{\cdot}35=3350$ kg/m²
Self weight of slab (25 cm) $=25\times24=600$ kg/m²
$w=3950$ kg/m²
Total load $w=3950$ kg/m²

16·4·1 Bending Moments Circumferential

$$M_{c_c}=\frac{wR^2}{16}=\frac{3950\times3{\cdot}3^2}{16}=2680\text{ mkg.}$$

16·4·2 Radial

$$M_{r_c}=\frac{wR^2}{16}=2680\text{ mkg. }(+)$$

$$M_{r_s}=\frac{2}{16}\,wR^2=2\times2680=5360\text{ mkg.}$$

16·4·3 Effective Depth

$$d_e=\sqrt{\frac{5360\times100}{100\times12{\cdot}6}}=\sqrt{425}=20{\cdot}6\text{ cm}$$

Use $d=25$ cm $d_e=21$ cm

16·4·4 Main Steel for Radial Moments

At support the steel is on the water face

$$A_{t_s}=\frac{5360\times100}{1000\times{\cdot}84\times21}=30{\cdot}5\text{ cm}^2$$

Use 20 mm ϕ bar at 10 cm c/c over a length $\left(R-\frac{R}{\sqrt{3}}\right)$

$$=3{\cdot}3-\frac{3{\cdot}3}{1{\cdot}73}=3{\cdot}3-1{\cdot}91=1{\cdot}39\ \text{m}$$

16·4·5 Main Steel for Circumferential and Radial Moment at Centre

Steel is away from water face,

$$A_{t_c}=\frac{2680\times 100}{1250\times 0{\cdot}864\times 21}=11{\cdot}9\ \text{cm}^2$$

Use 16 mm ϕ bar at 16·5 cm c/c

FIXED CIRCULAR SLAB

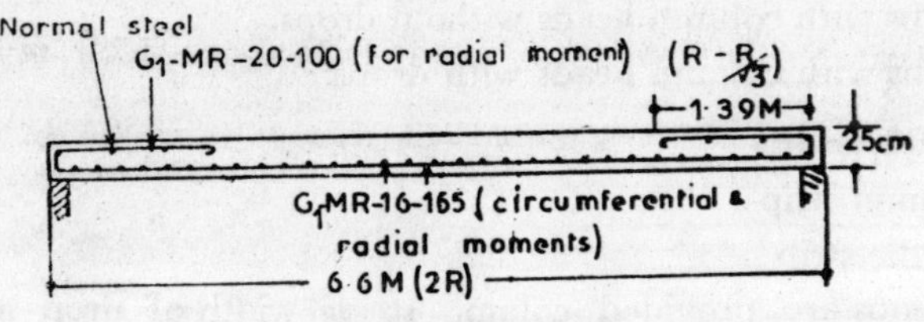

SECTION AA

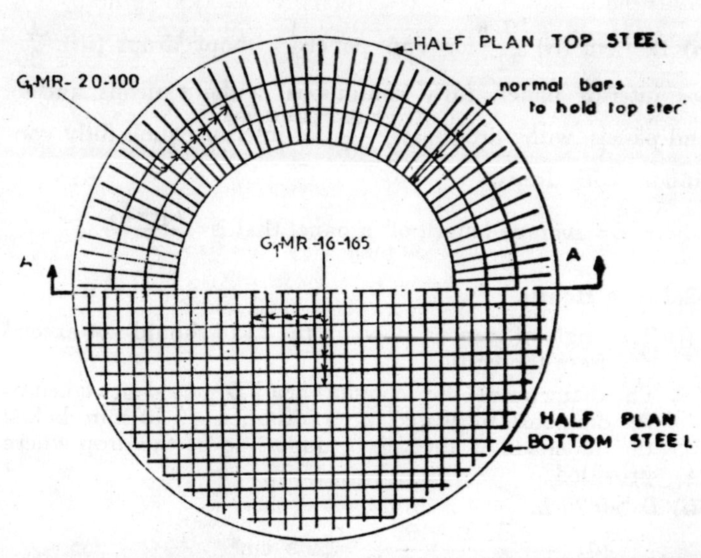

17

Flat Slabs–General

17·1 Definition

A reinforced concrete slab in which the load on the floor is carried directly to the columns without the use of beams is called a flat slab.

17·2 Types

(*i*) Slabs without column heads

(*ii*) Slabs with column heads without drops.

(*iii*) Slabs with column heads with drops.

17·3 Panel Divisions

(*i*) Column strip

(*ii*) Middle strip

where drops are provided column strip=width of drop and middle strip=width of panel—width of drop

17·4 Slab thickness

The slab thickness should not be less than the greatest of the following :

(*i*) 12·5 cm (*ii*) $\frac{L}{32}$ for end panels without drops (*iii*) $\frac{L}{36}$ for interior panel, fully continuous without drops and for end panels with drops (*iv*) $\frac{L}{40}$ for interior panels fully continuous with drops.

where L=average length of a panel that is=$\frac{L_1+L_2}{2}$

17·5 Column Heads

(*i*) The angle of greatest slope of the head should not exceed 45° from the vertical

(*ii*) The diameter of the column head D should be taken as its diameter measured at a distance of 40 mm below the underside of the slab or underside of the drop where provided.

(*iii*) $D \not> 0{\cdot}25\ L$.

(*iv*) Where the column and column head are not of circular cross—section, the diameter of the largest circle which can be drawn within the section, shall be taken as the column head diameter.

17·6 Design

(*i*) As Continuous Frames

In determining the bending moments and shearing forces the following assumptions may be made

(*a*) The structure may be considered to be divided longitudinally and transversely into frames consisting of a row of columns and strip of slab with a width equal to the distance between the centre lines of the panel on each side of the row of columns.

(*b*) Each Frame may be analysed by Hardy cross or other suitable methods. Each strip of floor and roof may be analysed as a separate frame with the columns above and below assumed fixed at their extremities. In the analysis the spans used should be the distance between the centres of support and when the slab is supported on a wall, distance from the face of the wall to the centre of column plus half the depth of the slab, should be taken as the slab.

17·6·1 Stiffness

The relative stiffness of the members either slab or column should be determined considering their cross sections of concrete in calculating the moment of inertia. The joints are assumed to have infinite moment of inertia. Variation of moment of inertia along the axis of the slabs should be taken into account.

17·6·2 Bending Moments

Bending moment at midspan and at the centre line of the supports should be calculated for the following conditions :

(*i*) Alternate spans loaded and all other spans unloaded.

(*ii*) Any two adjacent spans loaded and all other spans unloaded.

17·6·3 Design Moments

The maximum positive bending moment and the average negative bending moments, used in the design of any one span of the slab should, for the whole panel width, not be less than

$$M=\frac{wL_2}{10}\left(L_1-\frac{2D}{3}\right)^2$$

where

w=total load per unit area on the panel, D=diameter of column head, L_1=Length of the panel in the direction of span, L_2=width of the panel at right angle.

17·6·4 Bending Moment Distribution

Bending Moment Distribution % of M

	Column Strip	*Middle Strip*
Negative moment	75	25
Positive moment	55	45

Where column and middle strips are not equal in the case when column strip=width of drop and the middle strip is greater than ½ the panel width, the bending moments be increased in proportion to its increased width.

17·7 Empirical Method

This method is applicable when the following conditions are fulfilled :

(*i*) The slab should comprise a series of rectangular panels of approximately constant thickness arranged in at least three rows in two directions at right angles.

(*ii*) Ratio of the length of a panel to its width should not exceed 4 : 3.

(*iii*) The length or widths of any two adjacent panels in a series should not differ by more than 10% of the greater length or width.

(*iv*) End spans may be shorter but not longer than interior spans.

(*v*) Longer span should be taken in calculating the bending moments where adjacent spans differ.

(*vi*) The drops should be rectangular in plan and the length of drop in each direction should not be less than one third of the panel length.

(*vii*) For exterior panels the width of drop at right angles to the non-continuous edge and measured from the centre-line of the column should be equal to one half of the width of drop of interior panels.

17·7 Bending Moment Distribution

The bending moments are calculated using the expression and

$$M=\frac{wL_2}{10}\left(L_1-\frac{2}{3}D\right)^2$$

these are distributed between column and middle strips as given in the table.

17·7·2 Modified Moments

Where the width of the middle strip is greater than half the width of the panel, the moments in middle strip should be increased in proportion to its increased width. The moments in column strip be decreased by an amount such that there is no reduction in either the total positive or the total negative moments resisted by the column strip and middle strip together.

Panel details		Bending moment distribution % of M	
		Column strip	Middle strip
(*i*) Interior panels with drops			
Negative moments	...	50	15
Positive moments	...	20	15
(*ii*) Interior panels without drops			
Negative moments	...	46	16
Positive moments	...	22	16
(*iii*) Exterior panels with drops			
Exterior negative moment	...	45	10
Positive moment	...	25	19
Interior negative moment	...	50	15
(*iv* Exterior panels without drops			
Exterior negative moments	...	41	10
Positive moments	...	28	20
Interior negative moments	...	46	16

17·7·3 Shear Stresses

Shear stresses should not exceed the permissible shear at the critical sections as in Fig.

17·8 Reinforcement in two Directions

(*i*) The reinforeement should be so disposed that each strip is reinforced over its full width.

(*ii*) 40% of the +ve reinforcement should extend in the lower part of the slab to within a distance of 0·125 L from the line joining the centre of the column, in each strip.

(*iii*) The negative reinforcement in the top of the slab should extend into adjacent panels for an average distance measured from the line joining the centres of the columns of not less than 0·25 L and no bar should extend less than 0·2 L from this line.

(*iv*) The full area of negative reinforcement should be provided for a distance of not less than 0·2 L measured from the line joining the centres of the columns. The full area of positive reinforcement should be provided for a distance of not less than 0·25L measured from the panel.

(*v*) In the case of flat slabs supported on columns without heads or when the diameter is less than twice the average width of the top of the column, ⅔ of the amount of reinforcement required to resist the negative moment in the column strip should be placed in a width equal to half that of the coulmn strip and central with the column.

17·8·1 Reinforcement in four Directions

(*i*) The width of each direct band of reinforcement should be two-fifths of the panel width at right angles to the direction of the reinforcement.

(*ii*) The width of each diagonal band should be half the panel length or half the average panel length in the case of panels which are not squares.

(*iii*) (*a*) The reinforcement in the direct band should resist the entire positive moment on the column strip.

(*b*) The reinforcement in the diagonal bands should resist the entire positive moment on the middle strip.

(*c*) The reinforcement in the direct band plus the reinforcement in the diagonal bands should resist the negative moment on the column strip.

(*d*) Additional reinforcement should be provided to resist the negative moment on the middle strip.

(*iv*) The rules for slabs reinforced in two directions should be applied for direct band.

(*v*) 40% of the positive reinforcement should extend in the lower part of the slab to within a distance 0·2 L measured from a line through the centre of the column and at right angles to the direction of the band in each diagonal band.

(*vi*) The negative reinforcement in diagonal bands in the top of the slab should extend into adjacent panels for an average distance 0·4L beyond a line through the centre of the column and at right angles to the direction of the band and no bar should extend less than 0·35 L beyond this line.

(*vii*) In each diagonal band, full area of negative reinforcement should be provided for a distance of not less than 0·3 L measured from a line through the centre of the column and right angles to the direction of the band. The full area of positive reinforcement should be provided for a distance of not less than 0·35 L measured from a line through the centre of the panel and at right angles to the direction of the band.

(*viii*) The additional reinforcement required to resist the negative moment in the middle strip should extend for a distance of not less than 0·25L on each side of the line joining the centres of the columns.

(*ix*) For slabs with discontinuous edges, at all the discontinuous edges the positive and negative reinforcement should extend to within 75 mm of the edge of the panel and should be provided with U—hooks.

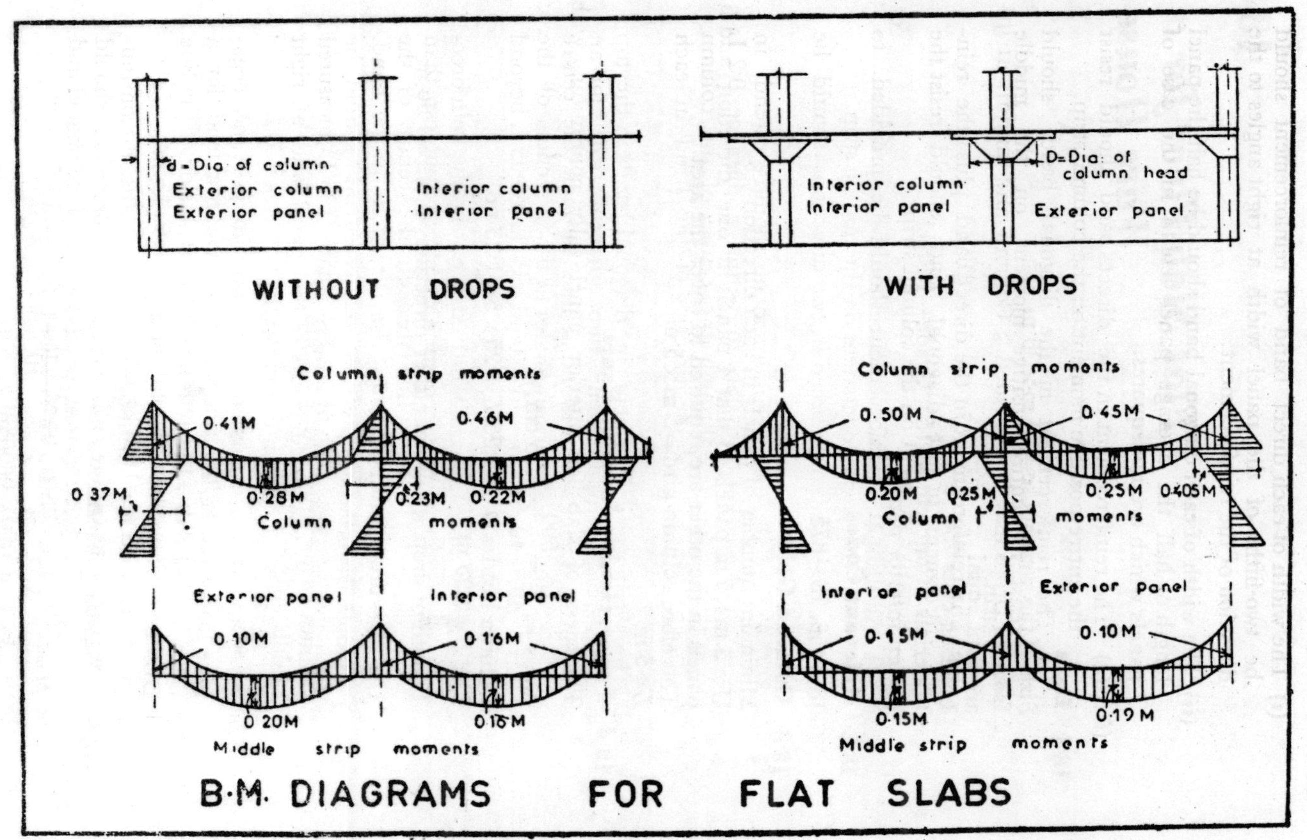

d-Dia of column
Exterior column
Exterior panel
Interior column
Interior panel
Interior column
Interior panel
D=Dia of column head
Exterior panel
WITHOUT DROPS
WITH DROPS
Column strip moments
0.41M
0.46M
0.37M
0.28M
0.23M
0.22M
Column moments
Column strip moments
0.50M
0.45M
0.20M
0.25M
0.25M
0.405M
Column moments
Exterior panel
Interior panel
0.10M
0.16M
0.20M
0.16M
Middle strip moments
Interior panel
Exterior panel
0.15M
0.10M
0.15M
0.19M
Middle strip moments
B.M. DIAGRAMS FOR FLAT SLABS

18

Flat Slab Floor for a Ware House

18·1 Data

Size of the werehouse=20 m×30 m
Isolated Drops may be provided. Column and middle strips are kept equal
Loading class 750
Materials : Concrete M 150, Grade—1 Steel
Design interior panel.

18·2 Relevant Codes

IS—456, IS—875

18·3 Column Grid

Minimum number of panels in each direction=3
Use 5 m×5 m panels so that 4 panels in one direction and 6 panels in the other are required to cover the area.
Therefore column gird=5 m×5 m
L=5 m

18·4 Proportioning Interior Panel

Thickness of slab for interior panel fully continuous with drops $=\frac{L}{40}=\frac{500}{40}=12{\cdot}5$ cm

Column head=D $\not> 0{\cdot}25\,L = {\cdot}25\times5=1{\cdot}25$ m
Use D=1.25 m

Length of drop $\not< \frac{L}{3}$ in either direction $\not<$ 5/3=1·66 m

Use D_w=drop width=2·5 m

18·4·1 Panel Divisions

Column strip=drop width=2 5 m
Middle strip=2·5 m

18·5 Loads

Live load=750 kg/m^2
Dead load of slab=12·5×24=300 kg/m^2
DL due to extra depth of slab at drops=50 kg/m^2
Total load w=1100 kg/m^2

18·5·1 Bending Moments

$$M=\frac{wL_2}{10}\left(L_1-\frac{2}{3}\,D\right)^2=\frac{1100\times5}{10}\left(5-\frac{2}{3}\times1{\cdot}25\right)^2$$
$$=550\times4{\cdot}165^2=9600 \text{ mkg}$$

18·5·2 Distribution of BM

For interior panel with drops.
Column strip bending moments
Negative $BM=50\%\ M=9600\times0{\cdot}5=4800$ mkg
Positive $BM=20\%\ M=9600\times0{\cdot}2=1920$ mkg
Middle strip Bending moments,
Negatige $BM=15\%\ M=0{\cdot}15\times9600=1440$ mkg
Positive $BM=15\%\ M=0{\cdot}15\times9600=1440$ mkg

18·5·3 Thickness of Slab

(*i*) Thickness of slab needed in drop area

$$d_d=\sqrt{\frac{4800\times100}{250\times8{\cdot}74}}=\sqrt{218}=14{\cdot}8 \text{ cm}$$

(*ii*) Thickness of slab needed in the area other than drops

$$d_s=\sqrt{\frac{1920\times100}{250\times8{\cdot}74}}=\sqrt{87{\cdot}5}=9{\cdot}3 \text{ cm}$$

Use $d_d=17{\cdot}5$ cm with effective depth$=15$ cm and $d_s=12{\cdot}5$ cm with effective depth$=10$ cm

18·5·4 Revised Loading

Increase in thickness of slab in drop area
$=17{\cdot}5-12{\cdot}5=5$ cm
Increase in DL$=5\times24=120$ kg/m^2
Total DL increase/panel$=4\times1{\cdot}25\times1{\cdot}25\times120=750$ kg

Increase in DL/m^2 of the entire panel$=\dfrac{750}{5\times5}=30$ kg/m^2

Actual total load
Live load$=750$ kg/m^2
DL of slab (12·5 cm)$=300$ kg/m^2
Increase in load due to drops$=30$ kg/m^2
Total load $w=1080$ kg/m$^2<1100$ kg/m^2 used for computing BM.
Total load/panel W$=1080\times25=27000$ kg

18·5·5 Shear Stress

Shear Stress at section $(D+d_d)$
Total load on the circular area with $(D+d_d)$ as diameter

$$W_1=\frac{\pi}{4}(D+d_d)^2\times w=\frac{\pi}{4}(1{\cdot}25+{\cdot}175)^2\times1080$$

$$=\frac{\pi}{4}\times1{\cdot}425^2\times1080=1670 \text{ kg.}$$

Shear Force=Total load load on the circular area
$W-W_1=27000-1670=25330$ kg.

$$\text{SF/metre of perimeter}=\frac{W-W_1}{\pi(D+d_d)}=\frac{25330}{\pi\times1{\cdot}425}=5780 \text{ kg}$$

$$\text{Shear stress}=\frac{5780}{100\times{\cdot}865\times15}=4{\cdot}42 \text{ kg/cm}^2<5 \text{ kg/cm}^2$$

Load acting on the square area $(D_w+d_s)=(D_w+d_s)^2\times w$
$=(2{\cdot}5+0{\cdot}125)^2\times1080=1080\times2{\cdot}625^2=7400$ kg.

Shear Force $=27000-7400=19600$ kg

Shear Force/metre of perimeter $=\dfrac{19600}{4\times 2\cdot 625}=1865$ kg

Shear stress $=\dfrac{1865}{100\times \cdot 865\times 10}=2\cdot 16$ kg/cm^2 <5kg/cm^2

18·5·6 Main Steel

Column strip

A_t for +ve moment $=\dfrac{1920\times 100}{1400\times 0\cdot 865\times 10}=15\cdot 8$ cm^2

18·6 Column Strip

A_t required/m width $=\dfrac{15\cdot 8}{2\cdot 5}=6\cdot 35$ cm^2

Use 10 mm ϕ bar at 12 cm c/c

A_t for −ve moment $=\dfrac{4800\times 100}{1400\times \cdot 865\times 15}=26\cdot 4$ cm^2

A_t/m width $=\dfrac{26\cdot 4}{2\cdot 5}=10\cdot 6$ cm^2

Use 16 mm ϕ bar at 18·5 cm c/c

18·7 Middle Strip

A_t needed for middle strip +ve and −ve moment

$=\dfrac{1440\times 100}{1400\times \cdot 865\times 10}=10\cdot 3$ cm^2

A_t/m width $=\dfrac{10\cdot 3}{2\cdot 5}=4\cdot 15$ cm^2

Use 10 mm ϕ bar at 18·5 cm c/c

Note: Please see 5 M×5 M grid drawings at the end of Chapter 19.

19

A Flat Slab Floor for a Ware House

19·1 Data

Column grid=8 m×8 m
Loading =Class—1000

Isolated drops are used. Column and middle strips are not equal

Materials : Concrete M 150, Grade—1 Steel

19·2 Relevant Codes

IS—456, IS—875

19·3 Proportioning Interior Panel

c/c of columns=L=8 m

Thickness of slab $\nless \frac{L}{40}$ =or 12·5 cm$=\frac{800}{40}=20$ cm

Column head $\ngtr 0{\cdot}25 \quad L=D=\frac{8}{4}=2$ m

Width of drop$=D_w \nless \frac{L}{3}$ in either direction $D_w=\frac{8}{3}$
$=2{\cdot}66$ m

Use $D_w=3$ m
D_w=width of drop=3 m
Column strip=Drop width=3 m
Middle strip=8−3=5 m

19·4 Loads

Live load=1000 kg/m²
DL of slab (20 cm)=20×24=480 kg/m²
DL due to extra thickness of slab in drops=120 kg/m²
Total load=w=1600 kg/m²

19·4·1 Bending Moment

$$M=\frac{wL_2}{10}\left(L-\frac{2}{3}\ D\right)^2=\frac{1600\times 8}{10}\left(8-\frac{2}{3}\times 2\right)^2$$

$=1280\times 6{\cdot}67^2=56500$ mkg.

19·4·2 Distribution of BM

For interior panels with drops when middle and column-strips are equal.

Column strip BM

Negative BM=50% M=56500×·5=28250 mkg

Positive BM=20% M=56500×·2=11300 mkg

Middle strip BM

Negative BM=15% M=56500×0·15=8475 mkg.

Positive BM=15% M=56500×0·15=8475 mkg.

19·4·3 Revised Moments

Since the column and middle strips are not equal, the moments should be revised.

When middle strip is equal to 4 m, Positive and Negative bending moments=8475 mkg.

Since middle strip is 5 m in width the moment should be increased proportionately.

Revised middle strip moment=$\frac{5}{4}$×8475=10,600 mkg.

Column strip moments.

Total column strip positive moment and middle strip negative moment=11300+8475=19775 mkg

Revised middle strip moment=10,600 mkg

Total column strip+ve moment and the middle strip

Revised moment=11300+10,600=21900 mkg

The amount of moment that can be decreased from column strip moments =21900−19775=2125 mkg

Column strip moments.

Positive moment=11300−2125=9175 mkg

Negative moment=28250−2125=26125 mkg

19·4·4 Thickness of Slab

(*i*) Thickness of slab at drops

$$d_a=\sqrt{\frac{26125\times100}{300\times8{\cdot}74}}=\sqrt{1000}=31{\cdot}5 \text{ cm}$$

(*ii*) Thickness of slab at middle strip

$$d_s=\sqrt{\frac{10600\times100}{500\times8{\cdot}74}}=\sqrt{243}=15{\cdot}6 \text{ cm}$$

(*iii*) Thickness of slab in the column strip in the area other than drop area

$$d_s=\sqrt{\frac{9175\times100}{300\times8{\cdot}74}}=\sqrt{350}=18{\cdot}7 \text{ cm}$$

Use d_a=33 cm with effective depth=31·5 cm in the drop area d_s=20 cm with effective depth=18·5 cm in the slab area.

19·4·5 Revised Loading

Increase in depth of drop slab=33−20=13 cm

Increase in dead load=13×24=312 kg/m²

Total increase per panel=$4\times1{\cdot}5\times1{\cdot}5\times312$=2800 kg

Increase in load/m² of panel=$\dfrac{2800}{8\times8}$=43·7 kg

Revised total load, live load =1000 kg/m²

Dead load due to slab (20 cm)=480 kg/m²

Due to drops =43·7 kg/m²

Total load =1523·7 kg/m²

Say w =1525 kg/m²

Which is less than 1600 kg/m² used for calculating BM

Total load/panel=W=$1525\times8\times8$=97600 kg

19·4·6 Shear Stresses

(*i*) Shear stress at section $D+d_d$

Total load on the circular area with $(D+d_d)$ as diameter.

$$W_1=\frac{\pi}{4}(D+d_d)^2w=\frac{\pi}{4}(2+0{\cdot}33)^2\times1525=6450 \text{ kg}$$

Shear force along the perimeter =97600−6450=91150 kg

Shear Force/m length of perimeter=$\dfrac{91150}{\pi\times2{\cdot}33}$=12447 kg

Shear stress=$\dfrac{12447}{100\times{\cdot}865\times31{\cdot}5}$=4·56 kg/cm²<5 kg/cm²

(*ii*) Shear stress at (D_w+d_s)

Load on the square area of $(D_w+d_s)=(D_w+d_s)^2w$

$=(3+0{\cdot}2)^2\times1525=3{\cdot}2^2\times1525=15300$ kg

Shear force along the perimeter=97600−15300

=82300 kg

Shear force/m length along perimeter=$\dfrac{82300}{3{\cdot}2\times4}$=642 9 kg

Shear stress=$\dfrac{6429}{100\times{\cdot}865\times18{\cdot}5}$=4·02 kg/cm²<5 kg/cm²

19·4·7 Main Steel

Column strip

A_t for positive moment=$\dfrac{9175\times100}{1400\times{\cdot}865\times18{\cdot}5}$=41 cm²

A_t/m width=$\dfrac{41}{3}$=13·66 cm²

Use 16mm ϕ bar at 14·5 cm c/c

A_t for Negative moment=$\dfrac{26125\times100}{1400\times{\cdot}865\times31{\cdot}5}$=68·5 cm²

$$A_t/\text{m width} = \frac{68{\cdot}5}{3} = 22{\cdot}83 \text{ cm}^2$$

Use 20mm ϕ bar at 13·5 cm c/c

Middle strip

$$A_t \text{ for } +\text{ve moment} = \frac{10600 \times 100}{1400 \times {\cdot}865 \times 18{\cdot}5} = 47{\cdot}4 \text{ cm}^2$$

$$A_t \text{ for } -\text{ve moment} = 47{\cdot}4 \text{ cm}^2$$

$$A_t/\text{m width} = \frac{47{\cdot}4}{5} = 9{\cdot}48 \text{ cm}^2$$

Use 12mm ϕ bar at 11·5 cm c/c for both positive and negative moments.

Note : Please see 8 M×8 M grid drawings on the back of 5 M×5 M grid drawings.

II. Design of Beams

1

Singly Reinforced Beams

1 Simply Supported Rectangular—General

1·1 Definition

A structural member supported without any fixity on load bearing walls or columns, subjected to roof or floor loads and reinforced on tension fibres only.

1·2 Limiting Stiffness

Span to depth ratio=20

1·3 Effective Depth

The distance between the centroid of tensile steel to the top compression fibre of concrete

$$d_e = d - d_t$$

1·3·1 Effective Span

(*i*) The effective span shall be the distance between the centres of supports.

(*ii*) Clear distance between supports plus the effective depth of beam whichever is smaller.

1·3·2 Minimum Reinforcement

Minimum tensile reinforcement in a beam shall not be less than 0·3 per cent of gross sectional area of the beam.

1·3·3 Cover for Reinforccment

(*i*) At each end of reinforcing bar not less than 25 mm, nor less than twice the diameter of the bar.

(*ii*) For longitudinal reinforcing bar, not less than 25 mm, nor less than the diameter of the bar.

(*iii*) For any other reinforcement not less than the diameter of such reinforcement.

1·3·4 Increased Cover

(*i*) Increased cover may be provided when concrete surface of members is exposed to the action of harmful chemicals. Increased cover may be between 15 mm and 40 mm in excess to the normal cover.

(*ii*) For concrete members immersed totally or periodically in sea water, the cover shall be 50 mm in excess to the normal cover specified.

1·3·5 Spacing of Reinforcement

(*i*) The horizontal distance between two parallel main reinforcement bars shall be not less than the greatest of the following :

(*a*) diameter of the bar if the diameters are equal.

(*b*) diameter of the larger bar if the diameters are unequal.

(*c*) 6 mm more than the nominal maximum size of the coarse aggregate used in the concrete.

(*d*) When needle vibrators are intended to be used the horizontal distance between bars of a group may be reduced to $\frac{2}{3}$ of the maximum size of the coarse aggregate.

(*e*) The minimum vertical distance between two horizontal main reinforcing bars shall normally be 15 mm, the maximum size of the coarse aggregate or maximum size of the bar whichever is greater.

1·3·6 Splice in Longitudinal Reinforcement

The length of lap in reinforcement shall not be less than

$$\frac{\text{bar diameter} \times \text{actual tensile stress}}{4 \times \text{permissible average bond stress}}$$

or 30 bar diameters whichever is greater.

1·4 Bending Moments

Bending moments in beams with free end supports shall be calculated for the effective span and for the loading thereon.

1·4·1 Shear Stresses

The shear stresses q at any cross section in a rectangular beam should be calculated from

$$q = \frac{Q}{bjd}$$

Q = total shear across the section

b = breadth of a rectangular beam

jd = lever arm of the resistsng moment.

1·4·2 Shear Stresses due to Torsional Moment

q' = Shear stress on the section resulting from torsional moment. M_t

$$= \frac{M_t\left(3+2\frac{b}{D}\right)}{b^2\,D}$$

where b and D are the width and depth of rectangular beam respectively.

(*i*) No shear reinforcement is necessary if $q+q' \ngtr$ permissible shear stress.

(*ii*) Provide full shear reinforcement to resist the whole shearing force due to transverse shear and the shear due to torsional moment if $q+q' >$ permissible shear stress.

(*iii*) If $q+q' > 4$ times the permissible shear stress, the beam should be redesigned.

1·4·3 Spacing of Shear Reinforcement Stirrups.

(*i*) The spacing of stirrups shall not exceed a distance equal to the lever arm of the resisting moment.

$$S=\frac{\sigma_{ss}\, A_w jd}{Q}$$

where S=spacing or pitch of stirrups.

A_w=total cross—sectional area of the stirrup legs effective in shear

jd=lever arm

Q=shear at the cross section

σ_{ss}=permissible tensile stress in the shear reinforcement=1400 kg/cm²

1·4·4 Bent up Bars as Shear Reinforcement

Total shear resistance Q of the bent up bar or bars

$$=\sigma_{ss}\, A_s \sin \alpha$$

where A_s=total cross—sectional area of the bent bar or bars

α=angle between the inclined bar and axis of the beam

1·4·5 Shear Reinforcement for Torsion

Reinforcement for resisting shear due to torsional moments shall be provided in the form of stirrups and longitudinal bars and shall be calculated from

$$A_w=\frac{M_t s}{0{\cdot}8\, \sigma_{ss}\, X_1\, Y_1}$$

where M_t=torsional moment

s=spacing

σ_{ss}=permissible tensile stress in the shear reinforcement

X_1, Y_1=sides of the rectangular stirrup

A_w=total cross-sectional area of the two legs of the rectangular stirrup.

1·5 Bond Stresses

Local bond stress$=\sigma_{bl}=\dfrac{\phi}{jd\, \Sigma\, 0}$

1·6 Useful Expressions

(*a*) From Equivalent stress diagram.

Critical neutral axis.

$$\frac{\sigma_{cb}}{\dfrac{\sigma_{st}}{m}}=\frac{n_c}{d_e-n_c}$$

or $$n_c = \frac{1}{1+\frac{\sigma_{st}}{m\sigma_{cb}}} d_e$$

(*b*) $C=T$

$$b \times n_e \times \frac{\sigma_{cb}}{2} = A_{st} \times \sigma_{st}$$

$$A_{st} = \frac{b \times n_c \; \sigma_{cb}}{2 \times \sigma_{st}}$$

(*c*) Taking moments about compressive force C

$$M = T \times jd_e = A_{st} \times \sigma_{cb} \times jd_e$$

Similarly

$$M = C \times jd_e = b \times n_e \times \frac{\sigma_{cb}}{2} \times jd_e = R \; b \; d_e^2$$

(*d*) Taking moments of compression and tension area, actual neutral axis

$$b \times n_a \times \frac{n_a}{2} = m \; A_{st} \; (d_e - n_a)$$

$$\frac{b \; n_a^2}{2} - m \; A_{st} \; n_a - m \; A_{st} \; d_e = 0$$

Solving the quadratic equation n_a can be obtained.

Balanced Section $n_c = n_a$

Under reinforced section $n_c > n_a$

Over reinforced section $n_c < n_a$

σcb KG/CM²	m	σst KG/CM²	R KG/CM²	n	j	pt
50	19	1000	10.22	0.488	0.838	1.22
		1300	9.07	0.422	0.859	0.81
		1400	8.74	0.404	0.865	0.72
		1900	7.41	0.333	0.888	0.44
		2100	6.97	0.311	0.896	0.37
		2300	6.59	0.292	0.903	0.32
		2600	6.23	0.274	0.909	0.26
70	13	1000	14.01	0.476	0.841	1.67
		1300	12.43	0.411	0.862	1.11
		1400	11.98	0.393	0.868	0.98
		1900	10.11	0.323	0.892	0.60
		2100	9.50	0.302	0.899	0.50
		2300	8.97	0.283	0.906	0.43
		2600	8.29	0.259	0.914	0.35

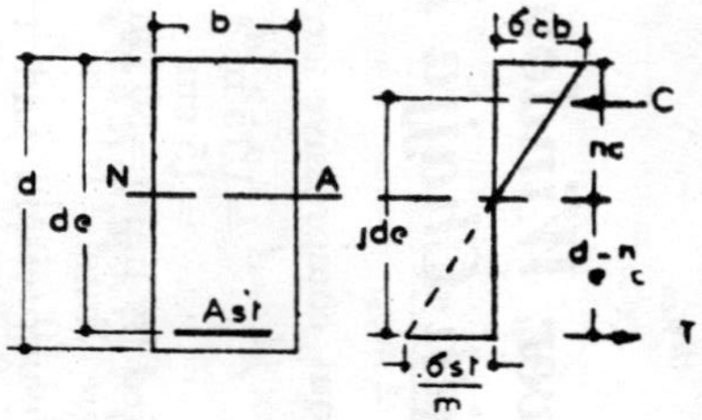

b, d, de, N, A, Ast, σcb, C, nc, jde, $d_e - n_c$, T, $\frac{\sigma st}{m}$

m = Modular ratio

$$= \frac{2800}{3\sigma cb}$$

$$pt = \frac{\sigma cb}{2\sigma st}\, n$$

$$R = \frac{1}{2}\sigma cb\, n\, j$$

$$M = Ast\ \sigma st\ j\,de = \frac{1}{2}\sigma cb\ n\ b\ j de$$

$$= R\, b\, de^2$$

COEFFICIENTS FOR SINGLY REINFORCED RECTANGULAR BEAMS

2

Lintel Beam over Window Opening with Chajja Projection

2·1 Data

Width of opening $=1{\cdot}15$ m
Bearing on either side $=15$ cm
Width of brick wall carried by lintel$=30$ cm
Height of brick wall above lintel$=1{\cdot}15$ m
Width of chajja$=\cdot 9$ m and 10 cm thick including plaster etc. ;
Materials : Concrete M 150, Grade—1 Steel.

2·2 Relevent Codes

IS—456, IS—875

2·2·1 Allowable Stresses

$C=50$ kg/cm^2
$t=1400$ kg/cm^2, $R=8{\cdot}74$, $a=\cdot 865\ d_e$

2·3·1 Effective Span

$l=1{\cdot}15+0{\cdot}15=1{\cdot}30$ m.

2·3 Depth of Beam

$d=$for simply supported beam$=\dfrac{l}{20}=\dfrac{130}{20}=6{\cdot}5$ cm

However use $d=20$ cm to suit to two layers of bricks

2·3·2 Width of Lintel

Width of lintel$=$width of masonary wall$=30$ cm

2·4 Loads

Height of masonary above lintel$=1{\cdot}15$ m
Height of the equilateral triangle$=\cdot 87l=\cdot 87\times 1{\cdot}3=1{\cdot}131$ m

Height of masonary wall carried by the lintel is approximately the same as of equilateral triangle.

Therefore entire brick wall load is taken.

Dead load of lintel$=\dfrac{20}{100}\times\dfrac{30}{100}\times 2400=144$ kg/m

Dead load from chajja$=\cdot 9\times 1\times\dfrac{10}{100}\times 2400=216$ kg/m

Live load from chajja$=\cdot 9\times 75=67{\cdot}5$ kg/m
Dead load due to brick wall (30 cm) 1·15 height$=1{\cdot}15\times 576$
$=662$ kg/m

Total load$=1089{\cdot}5$ kg/m, say 1090 kg/m

2·4·1 Bending Moment

$$\frac{wl^2}{8}=\frac{1090+1\cdot3^2}{8}=230 \text{ mkg}$$

$$d_e=\sqrt{\frac{23000}{30\times8\cdot74}}=\sqrt{87\cdot5}=9\cdot3 \text{ cm}$$

Use $d=20$ cm, $d_e=17$ cm

2·4·2 Main Steel

$$A_t=\frac{23000}{1400\times\cdot865\times17}=1\cdot12 \text{ cm}^2$$

Use 2—10 mm ϕ bars

2·4·3 Shear Stress

$$S_F=\frac{1090\times1\cdot3}{2}=708\cdot5 \text{ kg}$$

Shear stress $q_s=\dfrac{708\cdot5}{30\times\cdot865\times17}=1\cdot60 \text{ kg/cm}^2 < 5 \text{ kg/cm}^2$

Use nominal stirrups 6 mm ϕ at 15 cm c/c.

3

Lintel Beam over Door Opening

3.1 Data

Width of opening $=1{\cdot}45$ m
Bearing on either side $=15$ cm
Width of brick wall carried by lintel $=40$ cm
Height of brick wall above Lintel $=1{\cdot}15$ m
Materials : Concrete M 150, Grade—1 Steel.

3.2 Relevant Codes

IS—456, IS—875.

3.2.1 Allowable Stress

$C=50$ kg/cm^2
$t=1400$ kg/cm^2, $R=8{\cdot}74$, $a=0{\cdot}865\ d_e$

3.3 Effective Span

$l=1{\cdot}45+{\cdot}15=1{\cdot}60$ m

For simply supported beam$=\dfrac{l}{20}=\dfrac{160}{20}=8$ cm

Use $d=20$ cm to suit to two layers of bricks.

3.3.1 Width of Lintel

Width of lintel=masonary wall width=40 cm.

3.4 Loads

Height of masonary above lintel$=1{\cdot}15$ m
Height of equilateral triangle$={\cdot}87\ l={\cdot}87\times1{\cdot}6=1{\cdot}39$ m

Therefore height of masonary wall carried by the lintel is less than the height of equilateral triangle. The load of entire brick wall is taken

Dead load due to self weight of lintel$=\dfrac{20}{100}\times\dfrac{40}{100}\times2400$
$=192$ kg/m

Dead load due to brick wall above $=1{\cdot}15\times768$
$=882$ kg/m

Plastering etc : $=26$ kg/m

Total load w $=1100$ kg/m

3.4.1 Bending Moment

$$\frac{wl^2}{8}=\frac{1100\times1{\cdot}6^2}{8}=352 \text{ mkg}$$

3·4·2 Effective Depth

$$d_e=\sqrt{\frac{35200}{40\times 8{\cdot}74}}=\sqrt{101}=10{\cdot}1 \text{ cm}$$

Use $d=20$ cm, $d_e=17$ cm

3·4·3 Main Steel

$$A_t=\frac{35200}{1400\times{\cdot}865\times 17}=1{\cdot}71 \text{ cm}^2$$

Use 3—10 mm ϕ bars.

3·4·4 Shear Stress

$$S_F=\frac{1100\times 1{\cdot}6}{2}=880 \text{ kg.}$$

Shear stress $q_s=\dfrac{880}{40\times{\cdot}865\times 17}=1{\cdot}5 \text{ kg/cm}^2 < 5 \text{ kg/cm}^2$

Use nominal stirrups 6 mm ϕ—15 cm c/c.

4

Singly Reinforced Beam for a Residential Drawing Room

4·1 Data

Clear span = 5 m
Beam is supported on 30 cm brick wall and carries roof load.
Spacing of beams = 3 m
Slab thickness = 10 cm
Materials : Concrete M 150, Grade—1 Steel.

4·2 Relevant Codes

S—456, IS—875

4·2·1 Allowable Stresses

$C = 50$ kg/cm^2, $q_s = 5$ kg/cm^2, $q_b = 10$ kg/cm^2
$t = 1400$ kg/cm^2, $m = 19$, $R = 8{\cdot}74$, $a = {\cdot}865\, d_e$

4·3 Depth of Beam

For simply supported beams

$$d = \frac{\text{span}}{20} = \frac{530}{20} = 26{\cdot}5 \text{ cm}$$

Try 55 × 30 cm beam

$$d_e = 51{\cdot}5 \text{ cm.}$$

4·3·1 Effective Span

(*i*) Centre to centre of supports $= 5 + {\cdot}3 = 5{\cdot}3$ m
(*ii*) Clear span + effective depth $= 5 + {\cdot}515 = 5{\cdot}515$ m
$l = 5{\cdot}3$ m.

4·4 Loads

Live load (Roof) $= 3 \times 150 = 450$ kg/m

Dead load (10 cm) $= 3 \times \frac{10}{100} \times 2400 = 720$ kg/m

Roof finish (coba) $= 150 \times 3 = 450$ kg/m

Self weight of beam $= \frac{55}{100} \times \frac{30}{100} \times 2400 = 396$ kg/m

Total load $= w = 2016$ kg/m say $= 2020$ kg/m

4·4·1 Bending Moment

$$M = \frac{wl^2}{8} = \frac{2020 \times 5{\cdot}3^2}{8} = 7050 \text{ mkg.}$$

4·4·2 Shear Force

$$S=\frac{wl}{2}=\frac{2020\times5\cdot3}{2}=5360 \text{ kg}$$

4·4·3 Effective Depth

$$d_e=\sqrt{\frac{7050\times100}{30\times8\cdot74}}=\sqrt{2680}=51\cdot8 \text{ cm}$$

Use $d=55$ cm

Using 16 mm ϕ, $d_e=55-(2\cdot5+0\cdot8)=51\cdot9$ cm.

4·4·4 Main Steel

$$A_t=\frac{7050\times100}{1400\times\cdot865\times51\cdot9}=11\cdot21 \text{ cm}^2$$

Use 6—16 mm ϕ.

4·5 Shear Stress

$$\text{Shear stress}=\frac{S}{b\times a}=\frac{5360}{30\times\cdot865\times51\cdot9}$$
$$=4 \text{ kg/cm}^2 < 5 \text{ kg/cm}^2$$

No shear reinforcement is necessary.

However use nominal stirrups of 6 mm ϕ at a spacing equal to lever arm$=\cdot865\times51\cdot9=45$ cm c/c.

4·6 Bent up Bars

3 bars may be bent up out of 6 main bars

The distance from centre where these bars can be bent

$$x=\frac{l}{2}\sqrt{\frac{3}{6}}=\frac{5}{2}\times\cdot707=1\cdot76 \text{ m.}$$

Distance from the support$=2\cdot5-1\cdot76=\cdot74$ m$=74$ cm.

Using lattice girder theory the distance$=1\cdot414\times\cdot865\times51\cdot9$
$=63\cdot1$ cm

4·7 Bond Stress

Three bars are taken straight

$$q_b=\frac{5360}{\Sigma o\times a}=\frac{5360}{\cdot865\times51\cdot9\times3\times\pi\times1\cdot6}=7\cdot9 \text{ kg/cm}^2 < 10 \text{ kg/cm}^2$$

5

Singly Reinforced Beam to carry Class Room Floor

5·1 Data

Clear span = 7 m
Beam is supported on 40 cm brick walls
Centre to centre of beams = 2·75 m
Loading—class room floor
Slab thickness = 10 cm
Materials : Concrete M 150, Grade—1 Steel

5·2·1 Relevant Codes

IS—456, IS—875

5·2·2 Allowable Stresses

$C=50$ kg/cm², $q_s=5$ kg/cm², $q_b=10$ kg/cm²
$t=1400$ kg/cm², $m=19$, $R=8{\cdot}74$, $a={\cdot}865\ d_e$

5·3 Depth of Beam

For simply supported beams

$$d=\frac{\text{span}}{20}=\frac{740}{20}=37 \text{ cm}$$

Try $d=50$ cm, $d_e=45$ cm

5·3·1 Width of Beam

$b=25$ cm

5·3·2 Effective Span

Least of

(*i*) Centre to centre of supports $=7+0{\cdot}4=7{\cdot}4$ m
(*ii*) Clear span + effective depth $=7+{\cdot}45=7{\cdot}45$ m
$l=7{\cdot}4$ m

5·4 Loads

Live load $=2{\cdot}75\times400=1100$ kg/m

Dead load of slab $=2{\cdot}75\times\frac{10}{100}\times2400=660$ kg/m

Floor finish (IPS) $=2{\cdot}75\times60=165$ kg/m

Self weight of beam $=\frac{50}{100}\times\frac{25}{100}\times2400=300$ kg/m

Total load $w=2225$ kg/m

5·4·1 Bending Moment

$$M=\frac{wl^2}{8}=\frac{2225\times7{\cdot}4^2}{8}=15230 \text{ mkg}$$

5·4·2 Shear Force

$$S=\frac{wl}{2}=\frac{2225\times7{\cdot}4}{2}=8232{\cdot}5 \text{ say } 8240 \text{ kg.}$$

5·4·3 Effective Depth

$$d_e=\sqrt{\frac{15230\times100}{25\times8\cdot74}}=\sqrt{6970}=83\cdot4 \text{ cm}$$

$d=90$ cm and $d_e=85$ cm

Assumed trial section is not sufficient therefore revise the section

5·4·4 Depth of Section

$d=80$ cm
$b=35$ cm

5·5 Loads

$L.L.=1100$ kg/m, D.L. of slab=660 kg/m,
Floor finish=165 kg/m

$$\text{Self weight beam}=\frac{80}{100}\times\frac{35}{100}\times2400=672 \text{ say } 675 \text{ kg/m}$$

Total load$=w=2600$ kg/m

5·5·1 Bending Moment

$$M=\frac{wl^2}{8}=2600\times\frac{7\cdot4^2}{8}=17800 \text{ mkg}$$

5·5·2 Effective Depth

$$d_e=\sqrt{\frac{17800\times100}{35\times8\cdot74}}=\sqrt{5818}=76\cdot2 \text{ cm}$$

Use $d=80$ cm

Using 20 mm ϕ bars $d_e=80-\left(2\cdot5+\frac{2}{2}\right)=76\cdot5$ cm

5·5·3 Main Steel

$$A_t=\frac{17800\times100}{1400\times\cdot865\times76\cdot5}=19\cdot21 \text{ cm}^2$$

Use 7—20 cm ϕ in one layer

5·6 Shear Stress

$$\text{Shear force } S=\frac{wl}{2}=\frac{2600}{2}\times7\cdot4=9620 \text{ kg}$$

$$\text{shear stress}=\frac{S}{b\times a}=\frac{9620}{35\times\cdot865\times76\cdot5}=4\cdot15 \text{ kg/cm}^2$$

No shear reinforcement is necessary. However nominal strips of 6 mm ϕ at 66 cm c/c may be provided in addition to 3 bent up bars.

Distance from centre where three 20 mm ϕ bars can be bent up

$$x=\frac{l}{2}\sqrt{\frac{3}{7}}=\frac{7}{2}\times\cdot652=2\cdot28 \text{ m}$$

Distance from support$=3\cdot5-2\cdot28=1\cdot22$ m

Bend one bar at 122 cm from support and two bars at 122—65 =57 cm

Using lattice girder theory distance at which bars can be bent$=1\cdot414\times a=1\cdot414\times\cdot865\times76\cdot5=95\cdot5$ cm

5·7 Bond Stress

4 bars are taken straight

$$q_b = \frac{S}{\Sigma o \times a} = \frac{9620}{\cdot 865 \times 76{\cdot}5 \times 4 \times \pi \times 2} = 5{\cdot}8 \text{ kg/cm}^2 < 10 \text{ kg/cm}^2$$

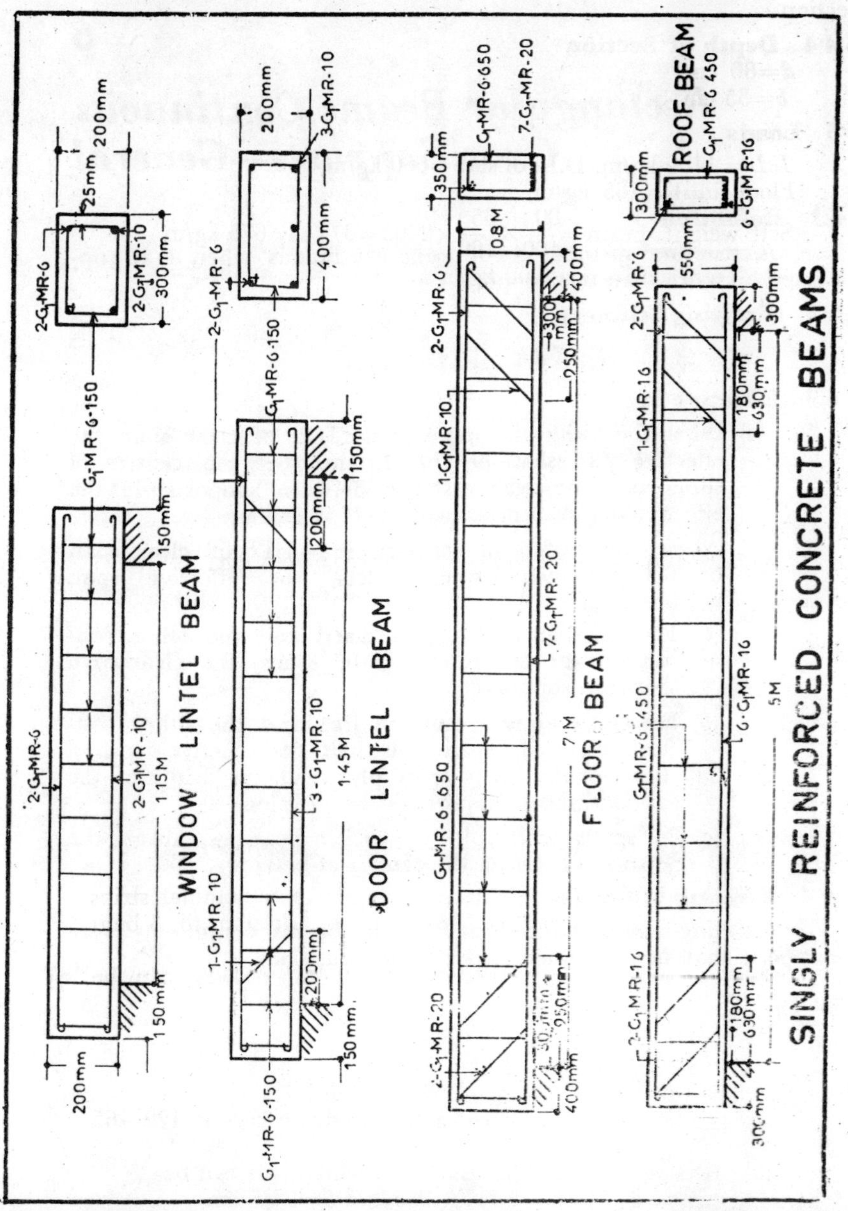

6

Rectangular Beams Continuous Over Supports—General

6·1 Definition

A structural member is said to be continuous when it is supported on more than two supports.

6·2 Limiting Stiffness

$$d \nless \frac{L}{25}$$

6·3 Effective Span

(*i*) When the width of supports is $<1/12$ of clear span, the effective span shall be the distance between centres of supports or the clear distance between supports plus the effective depth of beam whichever is smaller.

(*ii*) When the width of supports is $>1/12$ of clear span or 600 mm whichever is less, the effective span shall be :

(*a*) For end span with one end fixed and the other continuous or for intermediate spans, the clear span between supports.

(*b*) For end span with one end free and the other continuous, clear span plus half the effective depth of beam or the clear span plus half the width of the discontinuous support whichever is less.

(*c*) For spans with roller or rocker bearings, always the distance between the centres of bearings.

6·4 Bending Moments

Bending moments and shear forces in beams continuous over supports shall be calculated as per continuous slabs.

7

Continuous Corridor Beam

7·1 Data

Supports—stone pillars, 40 cm × 80 cm
Centre to centre of pillars = 2·75 m
Clear distance between pillars = 1·95 m
Clear width of corridor = 2·1 m
Centre to centre of corridor supports = 2·5 m
Height of 20 cm thick wall above the beam upto roof slab = 1·15 m
Slab thickness = 10 cm
10 cm brick parapet wall height above slab = 1 m
Materials : Concrete M 150, Grade—1 Steel.

7·2 Relevant Codes

IS—456, IS—875.

7·2·1 Allowable Stresses

$C = 50$ kg/cm², $q_s = 5$ kg/cm², $q_b = 10$ kg/cm²
$t = 1400$ kg/cm², $m = 19$, $R = 8{\cdot}74$, $a = {\cdot}865\, d_e$

7·3 Depth of Beam

For continuous beam

$$d = \frac{L}{25} = \frac{195}{25} = 7{\cdot}8 \text{ cm}$$

Assume $d = 20$ cm

7·3·1 Effective Span

Width of support = 80 cm

$$\frac{1}{12} \text{ clear span} = \frac{195}{12} = 16{\cdot}25 \text{ cm} < 80 \text{ cm}$$

Since the width of supports is wider than 1/12 span for interior span, the effective span = clear span between the supports,

$l = 1{\cdot}95$ m

7·4 Loads

Live loads/m length of beam
Live load from corridor slab (2·1 m wide)

$$= 400 \times \frac{2{\cdot}1}{2} = 420 \text{ kg/m}$$

Live load from Chajja (90 cm wide) at 150 kg/m²

$$= \frac{150 \times {\cdot}9}{1} = 135 \text{ kg/m}$$

Total Live load $= w_s = 555$ kg/m
Dead loads/m length of beam

Corridor slab (10 cm) 2·5 m wide$=240\times\frac{2\cdot5}{2}=300$ kg/m

Floor finish (2·5 cm) IPS$=60\times\frac{2\cdot1}{2}=63$ kg/m

Corridor parapet 10 cm brickwall 1 m height$=192$ kg/m

Brick wall above the beam to floor level (1·15 m) 20 cm thick brickwall $1\cdot15\times384=440$ kg/m

Self weight of beam$=\frac{20}{100}\times\frac{20}{100}\times2400=96$ kg/m

Chajja 90 cm wide (7·5 cm thick)$=\cdot9\times\frac{7\cdot5}{100}\times1\times2400$

$=162$ kg/m

Total dead load$=w_d=1253$ kg/m
Add for plaster, etc$=47$ kg/m
Total $w_d=1300$ kg/m

7·4·1 Bending Moments

$$=\frac{w_d l^2}{12}+\frac{w_s l^2}{9}=\frac{1300\times1\cdot95^2}{12}+\frac{555\times1\cdot95^2}{9}=411\cdot67+234\cdot33$$

$=646$ mkg

Central moment$=\frac{1300\times1\cdot95^2}{24}+\frac{555\times1\cdot95^2}{12}$

$=205\cdot83+176=381\cdot83$ mkg.

The support moment is greater than the moment at centre. Therefore 15% decrease in support moment and 15% increase in central moment is allowed.

7·4·2 Revised Moments

Support moment $=M_s=646-0\cdot15\times646=646-96\cdot9$
$=549\cdot1$ mkg
Central moment $=381\cdot83+96\cdot9=478\cdot73$ mkg
Maximum moment$=549\cdot1$ mkg.

7·4·3 Shear Force

Shear force at interior support

$$Q=0\cdot5\,W_d+0\cdot5\,W_s$$
$$=0\cdot5\times1300\times1\cdot95+0\cdot5\times555\times1\cdot95$$
$$=1267\cdot5+541\cdot1=1808\cdot6 \text{ kg}$$

7·4·4 Effective Depth

$$d_e=\sqrt{\frac{M}{b\times a}}=\sqrt{\frac{54910}{20\times8\cdot74}}=\sqrt{314}=17\cdot7 \text{ cm}$$

Use $d=20$ cm, $d_e=17$ cm.

Depth $d=20$ cm suits the height of one stone layer.

7·4·5 Main Steel

$$A_t=\frac{M}{t\times a}=\frac{54910}{1400\times\cdot865\times17}=2\cdot66 \text{ cm}^2$$

Use 4—10 mm ϕ (3·14 cm²) with two bent up bars.

Minimum steel required$=\frac{\cdot 3}{100}\times 20\times 20=1\cdot 2$ cm² $< 3\cdot 14$ cm².

7·4·6 Shear Stress

$$q_s=\frac{1808\cdot 6}{20\times \cdot 865\times 17}=6\cdot 14 \text{ kg/cm}^2 > 5 \text{ kg/cm}^2$$

7·5 Design of Stirrups

Therefore shear reinforcement is necessary
Twp bars are bent up to take shear
The distance from centre at which these bars can be bent up,

$$x=\frac{l}{2}\sqrt{\frac{A_b}{A_t}} \text{ where } A_b=\text{area of bent up bars}$$

A_t=total tensile steel

$$=\frac{195}{2}\sqrt{\frac{2}{4}}=\frac{195}{2}\times \cdot 707=68\cdot 2 \text{ cm.}$$

Distance from the edge of support$=\frac{195}{2}-68\cdot 2$

$$=97\cdot 5-68\cdot 2=29\cdot 3 \text{ cm.}$$

By using single system of lattice girder theory the distance from the support $=1\cdot 414\times a=1\cdot 414\times \cdot 865\times 17=20\cdot 7$ cm
Therefore bend 2 bars at 20 cm from edge of support.
Shear taken by bent up bars (45°)

$$Q=t\times A_t \sin \alpha=1400\times \frac{3\cdot 14}{2}\times \cdot 707=1550 \text{ kg} < 1808 \text{ kg}$$

The balance of shear to be resisted by stirrups
$=1808-1550=258$ kg.

The distance from support at which concrete alone takes shear

$$\frac{5}{6\cdot 15}=\frac{97\cdot 5-x}{97\cdot 5}$$

$$-\frac{5}{6\cdot 15}\times 97\cdot 5+97\cdot 5=x$$

$$x=-79+97\cdot 5$$
$$x=18\cdot 5 \text{ cm}$$

Shear stirrups are required from the support over a distance $=18\cdot 5$ cm

Spacing of 6 mm stirrups, $s=\frac{t\times A_w\times \cdot 865\, d_e}{Q}$.

$$=\frac{1400\times 2\times 0\cdot 28\times \cdot 865\times 17}{258}=43\cdot 98 \text{ cm say } 44 \text{ cm}$$

Maximum spacing allowed$=a=\cdot 865\times 17=14\cdot 7$ cm

Therefore use 6 mm—two legged stirrups at 14·5 cm c/c throughout.

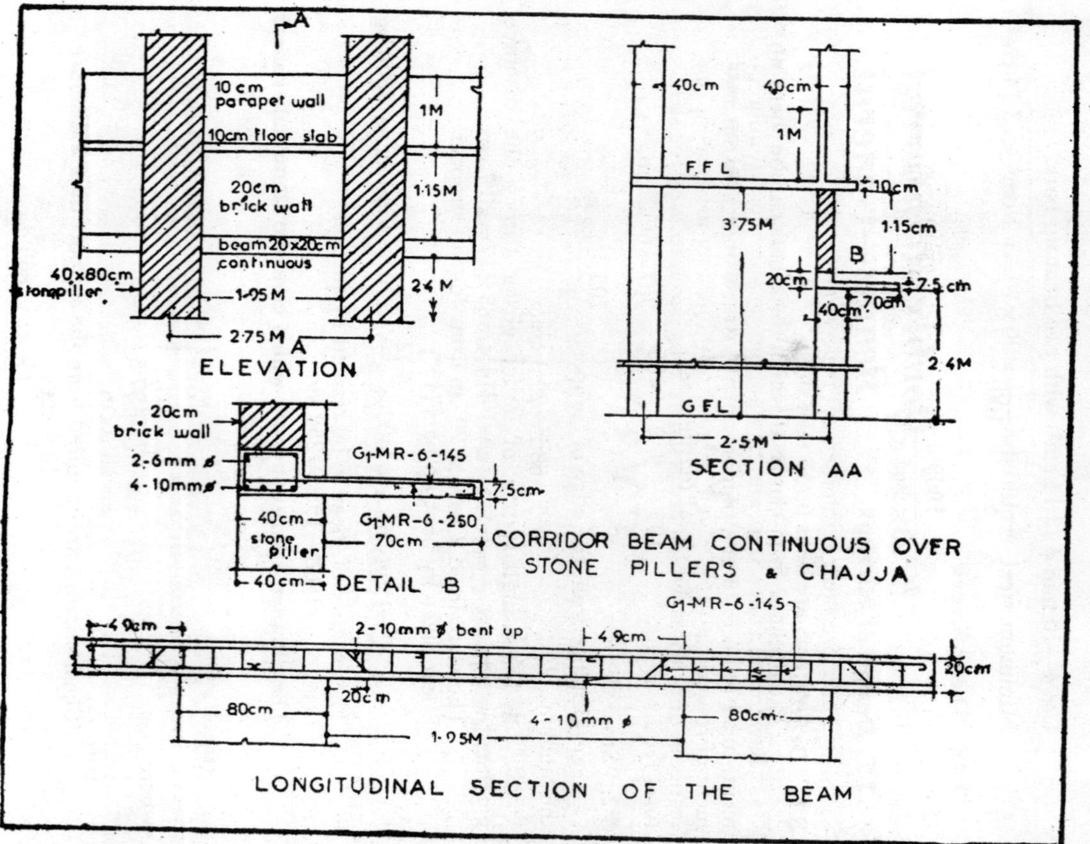
A
10 cm parapet wall
10cm floor slab
20cm brick wall
beam 20x20cm continuous
40x80cm stonepiller
1M
1·15M
2·4 M
1·95M
2·75M
A
ELEVATION
40cm
40cm
1M
F.F.L
10cm
3·75M
1·15cm
B
20cm
7·5 cm
70cm
40cm
2·4M
G.F.L
2·5M
SECTION AA
20cm brick wall
2-6mm ϕ
4-10mm ϕ
G1-MR-6-145
7·5cm
40cm
G1-MR-6-250
stone piller
70cm
40cm
DETAIL B
CORRIDOR BEAM CONTINUOUS OVER STONE PILLERS & CHAJJA
G1-MR-6-145
49cm
2-10mm ϕ bent up
49cm
20cm
20cm
80cm
4-10mm ϕ
80cm
1·95M
LONGITUDINAL SECTION OF THE BEAM

Doubly Reinforced Beams—General

8·1. Definition

Reinforced concrete beam containing compression reinforcement, in addition to tensile reinforcement, which increases the moment of resistance of a given section without any increase in the sectional dimensions which are restricted by architectural or structural requirements.

8·2. Steel Beams

If the amount of compressive reinforcement equals or exceeds the amount of tensile reinforcement, a doubly reinforced beam is designed as a steel beam by 'steel beam theory'.

8·3. Useful Expressions

(*i*) Moment of compression area

=Moment of tensile area

$$b \times n_a \times \frac{n_a}{2} - A_{sc}\,(n_a - d_c) + A_{sc}\,(n_a - d_c)$$

$$= m\,A_t\,(d - n_a)$$

$$b\,\frac{n^2{}_a}{2} + (m-1)\,A_{sc}\,(n_a - d_c) = m\,A_t\,(d - n_a)$$

(*ii*) $M_r = b\,n_a\,\frac{\sigma_{cb}}{2}\left(\frac{d - n_a}{3}\right) + (m-1)\,\sigma_{cb}'\,(d - d_c)$

where $\sigma_{cb}' = \sigma_{cb}\left(\frac{n_a - d_c}{n_a}\right)$

(*iii*) $M_r = A_t\,\sigma_{st}\,(d - d_c)$.

9

Doubly Reinforced Concrete Beam for a Residential Floor

9·1 Data

Clear Span=5m

Beam is supported on 30 cm brick walls.

Loading : Carries floor load in a residence.

Depth of beam is restricted to 50 cm.

Spacing of beams 3·5 m c/c.

Slab thickness=10 cm.

Materials : Concrete M 150, Grade—1 Steel

9·2 Relevant Codes

IS—456, IS—875.

9·2·1 Allowable Stresses

$C=50$ kg/cm², $q_s=5$ kg/cm², $q_b=10$ kg/cm².
$t=1400$ kg/cm², $R=8{\cdot}74$, $a={\cdot}865\ d_e$, $n_c={\cdot}404\ d_e$, $m=19$.

9·3 Effective Depth

Overall depth $d=50$ cm

Try $b=25$ cm

Using 20 cm dia. bars for tensile reinforcement in one layer

effective depth available$=d-\left(2{\cdot}5+\frac{2}{2}\right)=50-3{\cdot}5=46{\cdot}5$ cm.

9·3·1 Effective Span

(*i*) c/c of supports=5+0·3=5·30 cm.

(*ii*) Clear span+Effective depth=5+·465=5·465 m
$l=5{\cdot}3$ m.

9·4 Loads

Live load (floor of a residence)$=200\times3{\cdot}5=700$ kg/m

Dead load of slab (10 cm) $=10\times24\times3{\cdot}5=840$ kg/m

Floor finish at 60 kg/cm² $=60\times3{\cdot}5=210$ kg/m

Self weight of beam $={\cdot}50\times{\cdot}25\times2400=300$ kg/m

Total load$=w=2050$ kg/m.

9·4·1 Bending Moment

$$M=\frac{2050\times 5{\cdot}30^2}{8}=7200 \text{ mkg}$$

9·4·2 Shear Force

$$S=\frac{2050\times 5{\cdot}3}{2}=5400 \text{ kg}$$

9·4·3 Resisting Moment

Resisting moment of a balanced singly reinforced section

$$=M_1=8{\cdot}74\, bd^2_e=8{\cdot}74\times 25\times 46{\cdot}5^2=4740 \text{ mkg}$$

Remaining moment $M_2=M-M_1=7200-4740=2460$ mkg.

9·4·4 Tensile Steel

Tensile steel required for a singly reinforced Section to resist moment M_1

$$A_{t_1}=\frac{474000}{1400\times {\cdot}865\times 46{\cdot}5}=8{\cdot}4 \text{ cm}^2.$$

9·4·5 Additional Tensile Steel

Tensile steel required for remaining moment M_2

$$A_{t_2}=\frac{246000}{1400\times {\cdot}865\times 46{\cdot}5}=4{\cdot}36 \text{ cm}^2.$$

9·4·6 Compression Steel

Using 16 mm ϕ

d_c=cover for compression steel$=2{\cdot}5+\frac{1{\cdot}6}{2}=3{\cdot}3$ cm

Taking moments of additional tensile steel A_{t_2} and compression steel A_c

$$(m-1)\, A_c\, (n_c-d_c)=m\, A_{t_2}\, (d-n_c)$$

$$A_c=\frac{mA_{t_2}\,(d-n_c)}{(m-1)(n_c-d_c)}=\frac{19\times 4{\cdot}36\times(46{\cdot}5-0{\cdot}404\times 46{\cdot}5)}{(19-1)(0{\cdot}404\times 46{\cdot}5-3{\cdot}3)}$$

$$=\frac{19\times 4{\cdot}36\ (46{\cdot}5-18{\cdot}8)}{18\ (18{\cdot}8-3{\cdot}3)}$$

$$A_c=\frac{19\times 4{\cdot}36\times 27{\cdot}7}{13\times 15{\cdot}5}=8{\cdot}22 \text{ cm}^2.$$

Total tensile steel required$=A_{t_1}+A_{t_2}=8{\cdot}4+4{\cdot}36=12{\cdot}76$ cm²

This should not be less than $\cdot 3\%=\frac{\cdot 3}{100}\times 50\times 25=3{\cdot}76$ cm²

Use 5—20 cm ϕ

Compression steel required$=A_c=8{\cdot}22$ cm²

This should not be greater than 4%

Use 5—16 cm ϕ.

9·5 Shear Stresses

$$q_s=\frac{5400}{25\times 46{\cdot}5\times {\cdot}865}=5{\cdot}35 \text{ kg/cm}^2>5 \text{ kg/cm}^2$$

Hence shear reinforcement is necessary.

9·6 Bond Stress

If two 20 mm ϕ bars are bent up to take shear, remaining three bars will resist bond.

$$q_b=\frac{5400}{\cdot 865\times 46\cdot 5\times \pi\times 2\times 3}=7\cdot 11 \text{ kg/cm}^2<10 \text{ kg/cm}^2$$

9·7 Shear Reinforcement

Distance at which two bars can be bent

$$x=\frac{l}{2}\sqrt{\frac{2}{5}}=\frac{5\cdot 3}{2}\times\cdot 63=1\cdot 67 \text{ m}$$

$$\text{Distance from the support}=\frac{5}{2}-1\cdot 67=0\cdot 83 \text{ m}$$

If single system lattice girder theory is used, distance from support at which bars can be bent$=1\cdot 414\times\cdot 865\times 46\cdot 5=57$ cm

Shear taken by two bent up bars

$$Q=\sigma_{ss}\times A_s\times \sin\alpha=1400\times 2\times 3\cdot 14\times \sin 45°$$
$$=1400\times 6\cdot 28\times 0\cdot 707=6215\cdot 9 \text{ kg}>5400 \text{ kg}.$$

All the shear is taken up by the bent up bars.

Distance from the support upto which stirrups are required.

$$\frac{x}{2\cdot 5}=\frac{5}{5\cdot 35}$$

$$x=\frac{5\times 2\cdot 5}{5\cdot 35}=2\cdot 34 \text{ m}$$

$$2\cdot 5-2\cdot 34=\cdot 16 \text{ m}=16 \text{ cm}.$$

Over a distance of 16 cm stirrups are necessary. Use nominal stirrups 6 mm ϕ at $\cdot 865\times 46\cdot 5=40\cdot 2$ cm.

Say at 40 cm c/c throughout.

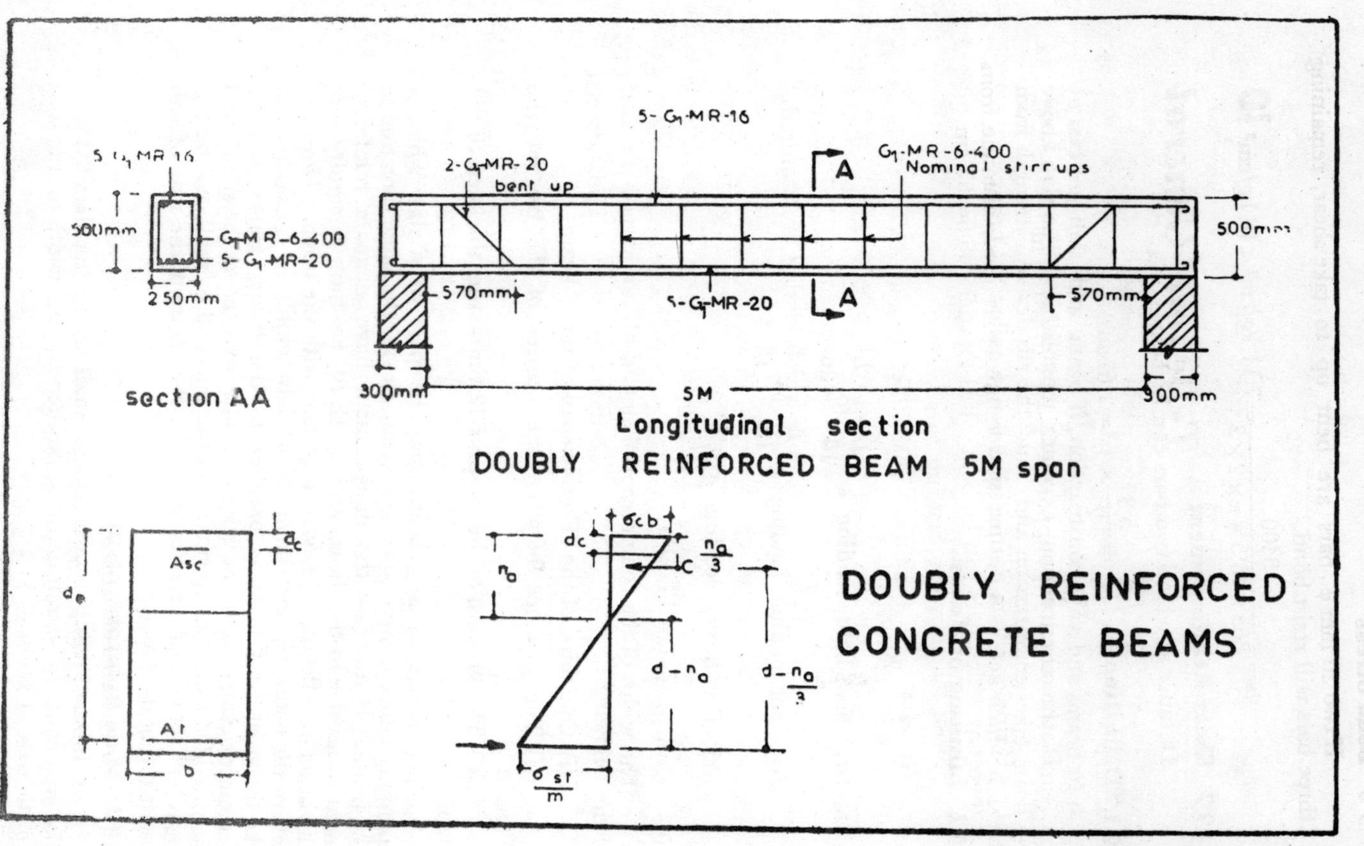
5 G1 MR 16
500mm
G1 MR-6-400
5- G1 MR-20
250mm
section AA
5- G1 MR-16
2-G1 MR-20
bent up
A
G1 MR-6-400
Nominal stirrups
500mm
570mm
5-G1 MR-20
A
570mm
300mm
5M
300mm
Longitudinal section
DOUBLY REINFORCED BEAM 5M span
Asc
de
dc
At
b
σcb
dc
na
C
na/3
d - na
d - na/3
σst/m
DOUBLY REINFORCED
CONCRETE BEAMS

10

T-Beams—General

10·1 Definition

In beams and slabs construction, if beam and slab are cast at one time, the construction becomes monolithic. The slab above the beam takes compression and thus the rib derives strength from the slab which acts as a compression flange called 'T' beam.

10·2 Limiting Stiffness

(*i*) $d \nless \frac{L}{20}$

(*ii*) For light loading $\nless \frac{L}{15}$ to $\frac{L}{20}$

(*iii*) For medium loading $\nless \frac{L}{12}$ to $\frac{L}{15}$

(*iv*) For heavy loading $\nless \frac{L}{12}$

10·3 Compression Flange

The width of the compression flange shall not exceed the least of the following :

(*i*) One third of the effective span of the T-beam

(*ii*) The distance between the centres of the ribs of the T-beam.

(*iii*) The breadth of the rib plus 12 times the thickness of the slab.

The flange of a T-beam may be part of a slab which is spanning either transverse to the beam or in the same direction as the beam. In any case the flange small have adequate reinforcement transverse to the beam and it shall be built integrally with the beam or effectively keyed together with the beam. However, where the main reinforcement of a slab which is considered as the flange of the T-beam is parallel to the beam, transverse reinforcement extending to the length of one quarter of the span, shall be provided near the top surface of the slab and it shall not be less than 60% of the main reinforcement in the centre of the span of slab constituting the flange.

10·4 Main Reinforcement

Minimum tensile reinforcement shall not be less than 0·3% of the area equal to overall depth multiplied by the width of the web in the case of T-beam.

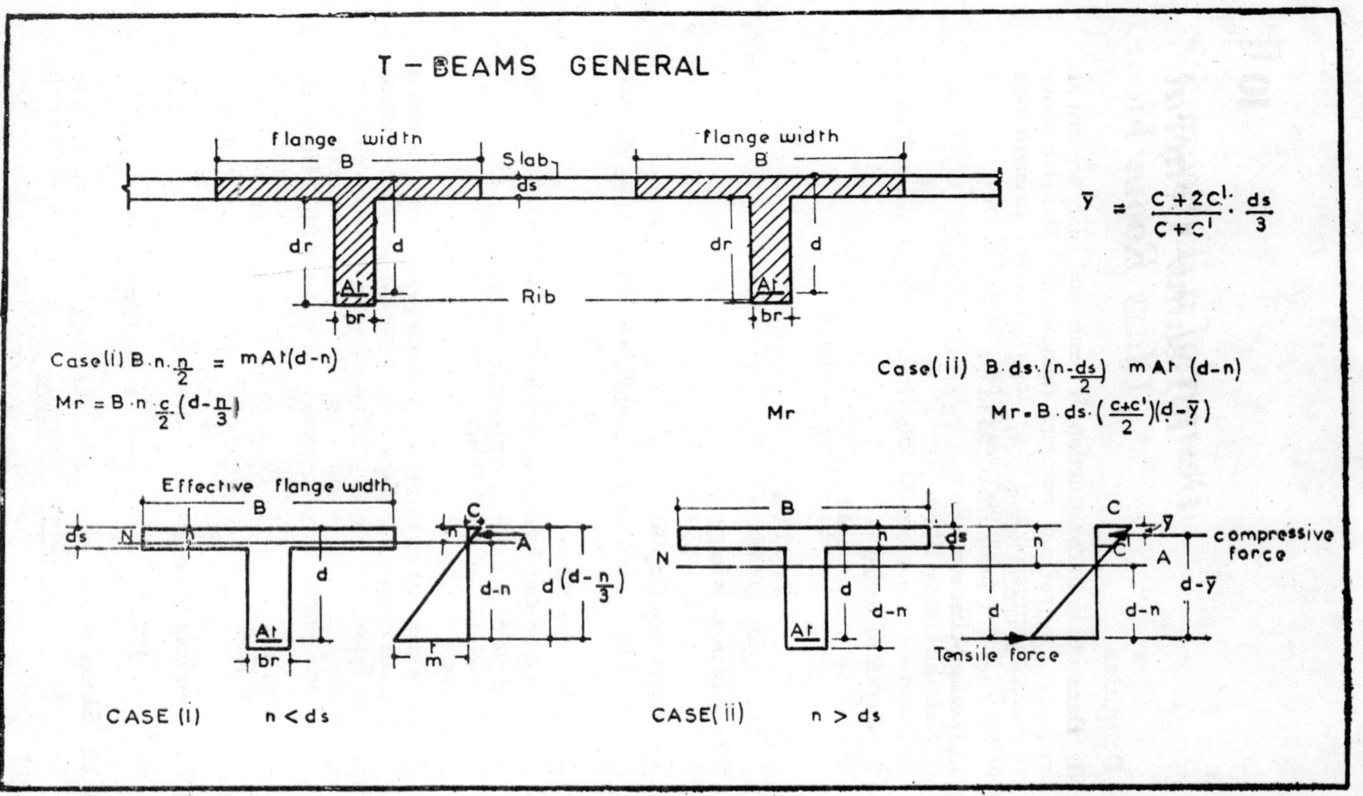
T - BEAMS GENERAL
flange width
B
Slab
ds
flange width
B
dr
d
At
Rib
br
$\bar{y} = \frac{C + 2C'}{C + C'} \cdot \frac{ds}{3}$
Case(I) $B \cdot n \cdot \frac{n}{2} = mAt(d-n)$
$Mr = B \cdot n \cdot \frac{c}{2}\left(d - \frac{n}{3}\right)$
Case(ii) $B \cdot ds \cdot \left(n - \frac{ds}{2}\right)$ $mAt\,(d-n)$
Mr
$Mr = B \cdot ds \cdot \left(\frac{c+c'}{2}\right)(d - \bar{y})$
Effective flange width
B
ds
N
C
n
A
d
d-n
$d\left(d - \frac{n}{3}\right)$
At
br
m
B
N
ds
n
d
d-n
At
C
ȳ
compressive force
C'
A
d-ȳ
Tensile force
CASE (I) n < ds
CASE(ii) n > ds

11

Design of a T-Beam for Class Room Floor

11·1 Data

Clear span = 7 m
Centre to centre of supports = 7·45 m
Spacing of T-beam rib = 2·75 m c/c
Loading-class room floor
Slab thickness = 10 cm
Materials : Concrete, M 150, Grade—1 Steel, Ribbed Trosteel

11·2 Depth of Beam

$$d = \frac{\text{Span}}{15} = \frac{745}{15} = 49\cdot6 \text{ m}$$

Try $d = 50$ cm
Rib depth = 40 cm

11·2·1 Width of Rib

$$b_r = 25 \text{ cm}$$

11·2·2 Effective Depth

Assuming in two layers of steel 25 mm dia e ctive depth

$$= 50 - (2\cdot5 + 2\cdot5 + 1) = 44 \text{ cm}$$

11.2·3 Effective Span

(*i*) Clear span + effective depth = 7 + ·44 = 7·44 m
(*ii*) Centre to centre of supports = 7·45 m,

$$l = 7\cdot44 \text{ m}$$

11·3 Loads

D L of slab (10 cm) = 240 × 2·75 = 660 kg/m
LL (class room) = 400 × 2·75 = 1100 kg/m
Floor finish = 60 × 2·75 = 165 kg/m
Rib self weight $= \frac{40}{100} \times \frac{25}{100} \times 1 \times 2400 = 240$ kg/m
Plaster etc = 2·75 × 50 = 137·5 kg/m
Total load = 2302·5 kg/m

Say $w = 2310$ kg/m

11·3·1 Bending Moment

$$BM = \frac{wl^2}{8} = \frac{2310 \times 7\cdot44^2}{8} = 16100 \text{ mkg}$$

11·3·2 Shear Force

$$S = \frac{2310 \times 7\cdot44}{2} = 8600 \text{ kg}$$

11·3·3 Main Steel

$$A_t=\frac{1610000}{1400\times\cdot9\times44}=29\cdot04\ \text{cm}^2$$

Use 6 Nos—25 mm ϕ, A_t provided—29·45 cm²
(or 10 Nos. of 20 mm ϕ) 31·42 cm²

11·4 Effective Flange Width

Least of

(*i*) c/c of beam $=275$ cm

(*ii*) 12 d_s+b_r $=12\times10+25=145$ cm

(*iii*) 1/3 effective span $=\frac{744}{3}=248$ cm

$$B=145\ \text{cm}$$

11·5 Stresses N.A.

$$B\times d_s\times\left(\frac{n-d_s}{2}\right)=m\times A_t\,(d-n)$$

$$145\times10\,(n-5)=19\times29\cdot45\,(44-n)$$

$$1450\,n-7250=24600-560\,n$$

$$2010\,n=31850$$

$$n=\frac{31850}{2010}=15\cdot8\ \text{cm}$$

c'=concrete stress at junction of rib and slab

$$=\left(\frac{n-d_s}{n}\right)c=\left(\frac{15\cdot8-10}{15\cdot8}\right)c=\frac{5\cdot8}{15\cdot8}\,c=\cdot368\,c$$

$$\bar{y}=\frac{c+2c'}{c+c'}\times\frac{d_s}{3}=\frac{1\cdot736}{1\cdot368}\times\frac{10}{3}=4\cdot23\ \text{cm}$$

Lever arm $d_e-\bar{y}=44-4\cdot23=39\cdot77$ cm

Stress in steel $=t=\frac{1610000}{39\cdot77\times29\cdot45}=1380\ \text{kg/cm}^2$

$$c=\frac{1380}{19}\times\frac{15\cdot8}{44-15\cdot8}=\frac{1380}{19}\times\frac{15\cdot8}{28\cdot2}=40\cdot6\ \text{kg/cm}^2$$

11·6 Shear Force

$$S=8600\ \text{kg}$$

$$q_s=\frac{8600}{25\times39\cdot77}=8\cdot64\ \text{kg/cm}^2$$

Hence shear reinforcement is required.
Section at which, concrete alone can take shear from centre

$$=\frac{5}{8\cdot64}\times3\cdot72=2\cdot13\ \text{m}$$

Therefore for 1·57 m from support should be provided.
Using lattice girder theory distance at which bars can be bent
$=1\cdot414\,a=1\cdot414\times39\cdot77=56$ cm
Shear carried by bent up bars$=0\cdot707\times A_w\times t_w$
$=0\cdot707\times3\times19\cdot63\times1400=58289\ \text{kg}>8600\ \text{kg}.$

Provide nominal stirrups 6 mm at 35 cm c/c throughout in addition to bent up bars.

11·7 Using Torsteel

$t=2300$ kg/cm²
$d=50$ cm rib depth=40 cm
Width of rib $=25$ cm

11·8 Trial Section

Using 20 mm ϕ, $d_e=50-(2{\cdot}5+2+1)=44{\cdot}5$ cm

11·9 Bending Moment

$$M=16100 \text{ mkg}$$
$$S=8600 \text{ kg}$$

11·9·1 Main Steel

$$A_t=\frac{161000}{2300\times{\cdot}9\times44{\cdot}5}=17{\cdot}4 \text{ cm}^2$$

Try 6—20 mm ϕ A_t provided=18·84 cm²

11·9·2 Stresses

$$B=145 \text{ cm}$$
$$B\times d_s\times\left(n-\frac{d_s}{2}\right)=m\times A_t\,(d-n)$$
$$145\times10\,(n-5)=19\times18{\cdot}84\,(44{\cdot}5-n)$$
$$1450\,n-7250=16000-359\,n$$
$$1809\,n=23250$$
$$n=\frac{23250}{1809}=12{\cdot}8 \text{ cm}$$

c'=concrete stress at junction of rib and slab,

$$=\left(\frac{n-d_s}{n}\right)c=\frac{(12{\cdot}8-10)c}{12{\cdot}8}=\frac{2{\cdot}8c}{12{\cdot}8}=0{\cdot}218\,c$$

$$\bar{y}=\frac{c+2c'}{c+c'}\times\frac{d_s}{3}=\frac{c+{\cdot}218c\times2}{c+{\cdot}218\,c}\times\frac{10}{3}=\frac{1{\cdot}436}{1{\cdot}218}\times\frac{10}{3}$$
$$=3{\cdot}9 \text{ cm}$$

$$a=de-\bar{y}=44{\cdot}5-3{\cdot}9=40{\cdot}6 \text{ cm}$$

$$\text{Stress in steel}=t=\frac{1610000}{40{\cdot}6-18{\cdot}84}=2100 \text{ kg/cm}^2<2300 \text{ kg/cm}^2$$

Stress in concrete

$$c=\frac{2100}{19}\times\frac{12{\cdot}8}{44{\cdot}5-12{\cdot}8}=\frac{2100}{19}\times\frac{12{\cdot}8}{31{\cdot}7}$$
$$=44{\cdot}5 \text{ kg/cm}^2<50 \text{ kg/cm}^2$$

11·10 Shear Stress

$$q_s=\frac{8600}{25\times40{\cdot}6}=8{\cdot}45 \text{ kg/cm}^2>5 \text{ kg/cm}^2$$

Hence shear reinforcement is required. Use the arrangement as per mild steel.

12

T-Beam for Class Room Floor Carrying 3·75 m High Partition Wall

12·1 Data

Clear span $=7$ m
Centre to centre of supports $=7\cdot45$ m
Spacing of ribs of T-beam $=2\cdot75$ m c/c
Slab thickness $=10$ cm
Materials : Concrete M 150, Grade—1 Steel, Ribbed Torsteel

12·2 Depth

$$\text{Depth of beam} = \frac{\text{span}}{15} = \frac{745}{15} = 49 \text{ cm}$$

Try $d=50$ cm
Rib depth $=40$ cm

12·2·1 Width of Rib

$b_r=25$ cm Say

12·2·2 Effective Depth

Assuming two layers of steel, 25 mm bars effective depth

$$d_e=50-(2\cdot5+2\cdot5+1)=44 \text{ cm}$$

12·2·3 Effective Span

(*i*) Clear span+effective depth $=7+\cdot44=7\cdot44$ m
(*ii*) Centre to centre of supports $=7\cdot45$ m
$l=7\cdot44$ m

12·3 Loads

Dead load of slab (10 cm) $=240\times2\cdot75=660$ kg/m
Live load $=400\times2\cdot75=1100$ kg/m
Floor finish $=2\cdot75\times60=165$ kg/m

Self weight of rib $=\dfrac{40}{100}\times\dfrac{25}{100}\times1\times2400=240$ kg/m

Plaster on ceiling, beam etc $=2\cdot75\times50=137$ kg/m
Partition wall at 192 kg/m^2 $=192\times3\cdot75=720$ kg/m
Total load on beam $=3022$ kg/m Say $w=3025$ kg/m

12·3·1 Bending Moment

$$\text{BM}=\frac{wl^2}{8}=\frac{3025\times7\cdot44^2}{8}=21100 \text{ mkg}$$

12·3·2 Shear Force

$$S=\frac{3025\times 7{\cdot}44}{2}=11300 \text{ kg}$$

12·3·3 Main Steel

$$A_t=\frac{21100\times 100}{1400\times{\cdot}9\times 44}=38{\cdot}2 \text{ cm}^2$$

8 No. bars 25 mm ϕ

A_t provided$=39{\cdot}27$ cm^2

12·4 Effective Flange Width

Least of

(*i*) c/c of beam$=2{\cdot}75$ m$=275$ cm

(*ii*) $12d_s+b_r \quad =12\times 10+25=145$ cm

(*iii*) $\frac{1}{3}$ effective span$=\frac{744}{3}=248$ cm

$$B=145 \text{ cm}$$

12·5 Stresses N.A.

$$B\times d_s\times\left(n-\frac{d_s}{2}\right)=m\times A_t\,(d-n)$$

$$145\times 10(n-5) \qquad =19\times 39{\cdot}27\,(44-n)$$

$$1450n-1450\times 5=19\times 39{\cdot}27\times 44-19\times 39{\cdot}27\times n$$

$$1450n-7250 \qquad =33000-749n$$

$$2199n \quad =40250$$

$$n=\frac{40250}{2199}=18{\cdot}5 \text{ cm}$$

c'=concrete stress at junction of rib and slab

$$=\left(\frac{n-d_s}{n}\right)c=\left(\frac{18{\cdot}5-10}{18{\cdot}5}\right)={\cdot}46c$$

$$\bar{y}=\frac{c+2c'}{c+c'}\times\frac{d_s}{3}$$

$$=\frac{c+2\times 0{\cdot}46c}{c+{\cdot}46c}\times\frac{10}{3}=\frac{1{\cdot}96c}{1{\cdot}46c}\times\frac{10}{3}=4{\cdot}48 \text{ cm}$$

Lever arm$=d_c-\bar{y}=44-4{\cdot}48=39{\cdot}52$ cm

$$\text{Stress in steel}=t=\frac{2110000}{39{\cdot}52\times 39{\cdot}27}=1361 \text{ kg/cm}^2$$

$$\text{Stress in concrete}=c=\frac{1361}{19}\left(\frac{18{\cdot}5}{44-18{\cdot}5}\right)$$

$$=\frac{1361}{19}\times\frac{18{\cdot}5}{25{\cdot}5}=52 \text{ kg/cm}^2\simeq 50 \text{ kg/cm}^2$$

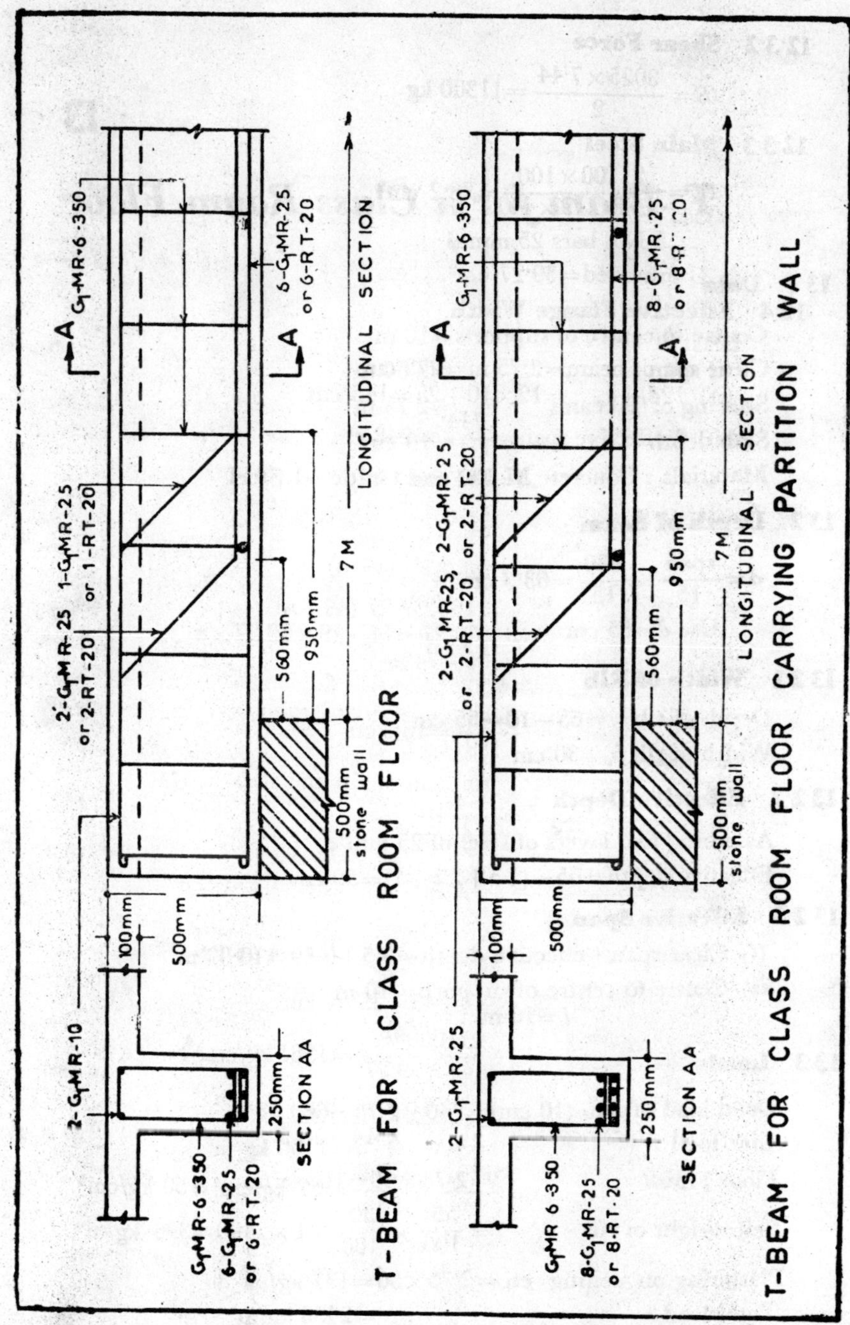
2-G1MR-10
G1MR-6-350
6-G1MR-25
or 6-RT-20
100mm
500mm
250mm
SECTION AA
2'-G1MR-25
or 2-RT-20
1-G1MR-25
or 1-RT-20
G1MR-6-350
6-G1MR-25
or 6-RT-20
A
A
500mm
stone wall
560mm
950mm
7M
LONGITUDINAL SECTION
T-BEAM FOR CLASS ROOM FLOOR
2-G1MR-25
G1MR 6-350
8-G1MR-25
or 8-RT-20
100mm
500mm
250mm
SECTION AA
2-G1MR-25,
or 2-RT-20
2-G1MR-25
or 2-RT-20
G1MR-6-350
8-G1MR-25
or 8-RT-20
A
A
500mm
stone wall
560mm
950mm
7M
LONGITUDINAL SECTION
T-BEAM FOR CLASS ROOM FLOOR CARRYING PARTITION WALL

13

T–Beam for a Class Room Floor

13·1 Data

Centre to centre of supports = 10 m
Clear span = 9·50 m
Spacing of T-beams = 2·75 m
Slab depth = 10 cm
Materials : Concrete M 150 and Grade—1 Steel

13·2 Depth of Beam

$$d = \frac{\text{span}}{15} = \frac{950}{15} = 63{\cdot}3 \text{ cm}$$

Use $d = 65$ cm

13·2·1 Width of Rib

Depth of rib = 65 − 10 = 55 cm
Width of rib $b_r = 30$ cm

13·2·2 Effective Depth

Assuming two layers of steel of 25 mm ϕ
Effective depth = 65 − (2·5 + 2·5 + 1) = 59 cm

13·2·3 Effective Span

(*i*) Clear span + effective depth = 9·5 + 0·59 = 10·09 m
(*ii*) Centre to centre of supports = 10 m
$l = 10$ m

13·3 Loads

Dead load of slab (10 cm) = 240 × 2·75 = 660 kg/m
Live load = 400 × 2·75 = 1100 kg/m
Floor Finish = 2·75 × 60 = 165 kg/m

Self weight of rib $= \frac{55}{100} \times \frac{30}{100} \times 1 \times 2400 = 396$ kg/m

Plastering on ceiling etc = 2·75 × 50 = 137 kg/m
Total load = 2458 kg/m
Say $w = 2500$ kg/m

13·3·1 Bending Moment

$$BM=\frac{wl^2}{8}=\frac{2500\times 10^2}{8}=31250 \text{ mkg}$$

13·3·2 Shear Force

$$S=\frac{2500\times 10}{2}=12500 \text{ kg}$$

13·3·3 Main Steel

$$A_t=\frac{31250\times 100}{1400\times \cdot 9\times 59}=42 \text{ cm}^2$$

9—25 mm ϕ

A_t provided$=44{\cdot}18$ cm²

13·4 Effective Flange Width

Least of

(*i*) c/c of beam $=275$ cm

(*ii*) $12d_s+b_r=12\times 10+30=150$ cm

(*iii*) ⅓ effective span$=\frac{10\times 100}{3}=333$ cm

$B=150$ cm

13·5 Stresses N.A.

$$B\times d_s\times\left(n-\frac{d_s}{2}\right)=m\times A_t\ (d-n)$$

$$150\times 10\times(n-5)\ =19\times 44{\cdot}18\ (59-n)$$

$$1500\ n-7500\ =49600-840\ n$$

$$2340\ n=57100$$

$$n=\frac{57100}{2340}=24{\cdot}5 \text{ cm}$$

c'=concrete stress at the junction of rib and slab

$$=\left(\frac{n-d_s}{n}\right)c=\left(\frac{24{\cdot}5-10}{24{\cdot}5}\right)c=\frac{14{\cdot}5}{24{\cdot}5}\times c=0{\cdot}592\ c$$

$$\bar{y}=\frac{c+2c'}{c+c'}\times\frac{d_s}{3}=\frac{c+2\times\cdot 592c}{c+\cdot 592c}\times\frac{10}{3}=\frac{2{\cdot}184}{1{\cdot}592}\times\frac{10}{3}=4{\cdot}56 \text{ cm}$$

Lever arm$=d_s-\bar{y}=59-4{\cdot}56=54{\cdot}44$ cm

$$\text{Stress in steel}=t=\frac{31250\times 100}{54{\cdot}44\times 44{\cdot}18}=1300 \text{ kg/cm}^2$$

$$\text{Stress in concrete}=c=\frac{1300}{19}\left(\frac{24{\cdot}5}{59-24{\cdot}5}\right)$$

$$=\frac{1300}{19}\left(\frac{24{\cdot}5}{34{\cdot}5}\right)=48{\cdot}5 \text{ kg/cm}^2$$

13·6 Shear Reinforcement

$S=12500$ kg

Shear stress$=\frac{S}{b\times a_{\infty}}=\frac{12500}{30\times 54\cdot 44}=7\cdot 65$ kg/cm^2 > 5 kg/cm^2

Hence shear reinforcment is required. Section upto which concrete can resist shear from centre$=\frac{5}{7\cdot 65}\times 5=3.26$ m

Shear reinforcement is required over a distance of 1·74 m from support

Using lattice girder theory distance at which bars can be bent $=1\cdot 414\times a=1\cdot 414\times 54\cdot 44=77$ cm.

Distance at which 4 bars can be bent up$=\frac{l}{2}\sqrt{\frac{4}{9}}$

$=\frac{10}{2}\times\cdot 665=3\cdot 3$ m from centre

From support the distance$=5-3\cdot 3=1\cdot 7$ m

4 bars bent only at 131 cm are therefore alright.

Shear carried by bent up bars$=0\cdot 707\times A_w\times t_w$

$=0\cdot 707\times 4\times 19\cdot 63\times 1400=78000$ kg > 12500 kg

Provide nominal stirrups. 6 mm at 50 cm c/c throughout in addition to bent up bars.

13·7 Using Torsteel

$t=2300$ kg/cm^2

13·8 Trial Section

$d=65$ cm, depth of rib$=55$ cm

width of rib$=30$ cm

13·8·1 Effective Depth

Using 20 mm ϕ bar

$d_e=65-(2\cdot 5+2+1)=59\cdot 5$ cm

13·8·2 Bending Moment

$M=31250$ mkg

Shear Force

$S=12500$ kg

13·8·3 Main Steel

$$A_t=\frac{3125000}{2300\times\cdot 9\times 59\cdot 5}=25\cdot 5\ \text{cm}^2$$

Try 9—20 mm ϕ . A_t provided$=28\cdot 28$ cm^2

13·8·4 Effective Flange Width

$B=150$ cm

13·9 Stresses

$$B\times d_s\left(n-\frac{d_s}{2}\right)=m\times A_t(d-n)$$

$$150\times 10(n-5)=19\times 28\cdot 28(59\cdot 5-n)$$

$$1500\,n-7500=32000-540\,n$$

$$2040\,n=39500$$

$$n=\frac{39500}{2040}=19{\cdot}3\text{ cm}$$

c' = concrete stress at the junction of rib and slab.

$$=\left(\frac{n-d_s}{n}\right)c=\left(\frac{19{\cdot}3-10}{19{\cdot}3}\right)c$$

$$=0{\cdot}478\,c$$

$$\bar{y}=\frac{c+2\,c'}{c+c'}\times\frac{d_s}{3}=\frac{c+2\times{\cdot}478\,c}{c+{\cdot}478\,c}\times\frac{10}{3}=\frac{1{\cdot}956}{1{\cdot}478}\times\frac{10}{3}=4{\cdot}45\text{ cm}$$

$$a=d_e-\bar{y}=59{\cdot}5-4{\cdot}45=55{\cdot}05\text{ cm}$$

$$\text{Stress in steel}=t=\frac{31250\times100}{55{\cdot}05\times28{\cdot}28}=1995\text{ kg/cm}^2<2300\text{ kg/cm}^2$$

$$c=\frac{1995}{19}\left(\frac{19{\cdot}3}{59{\cdot}5-19{\cdot}3}\right)=\frac{1995}{19}\times\frac{19{\cdot}3}{40{\cdot}2}$$

$$=50{\cdot}2\text{ kg/cm}^2\simeq50\text{ kg/cm}^2$$

13·10 Shear Stress

$$q_s=\frac{12500}{30\times55{\cdot}05}=7{\cdot}5\text{ kg/cm}^2$$

Hence use the shear reinforcement as per mild steel.

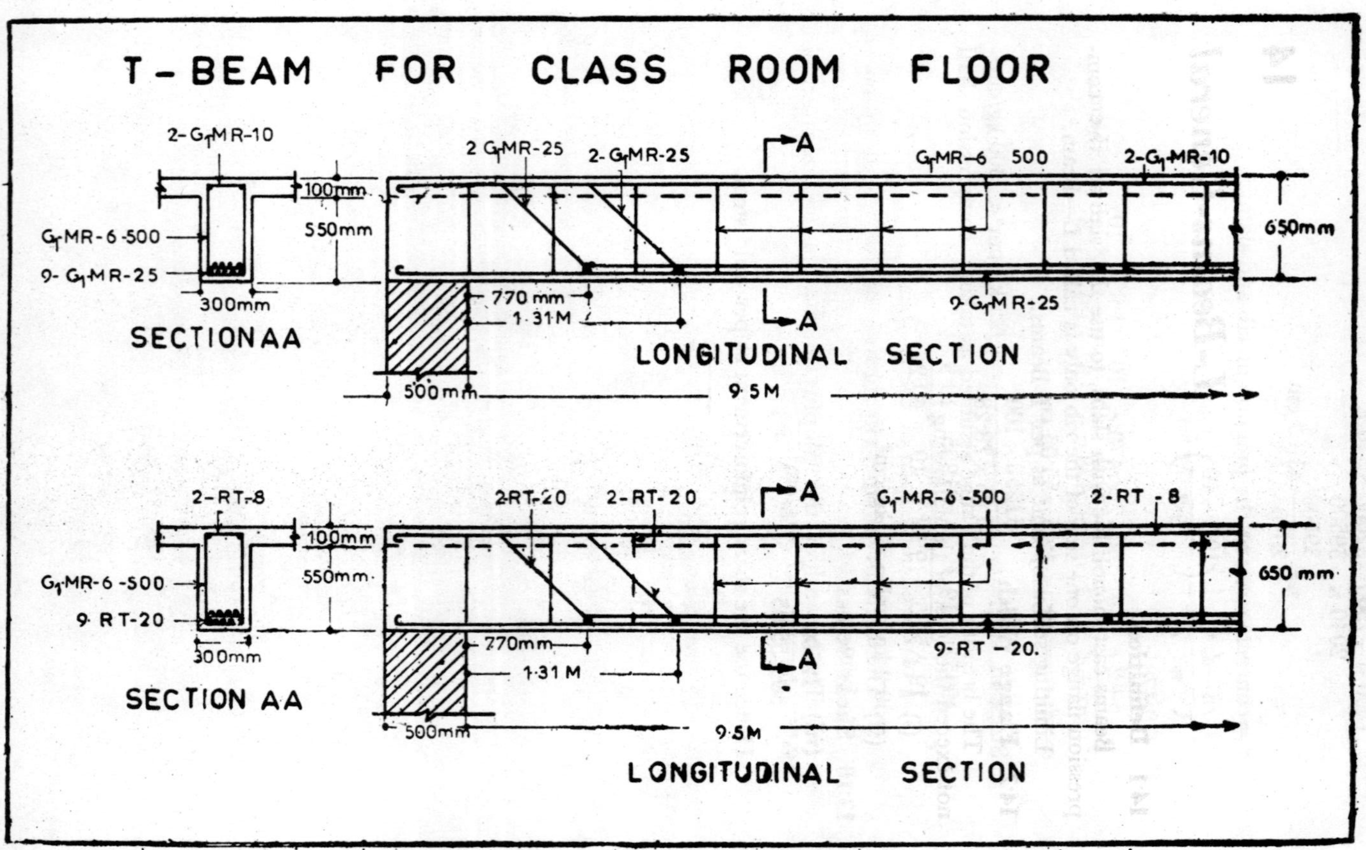
T - BEAM FOR CLASS ROOM FLOOR
2-G1MR-10
G1MR-6-500
9- G1MR-25
100mm
550mm
300mm
SECTIONAA
2 G1MR-25
2-G1MR-25
A
G1MR-6 500
2-G1-MR-10
650mm
770 mm
1·31 M
A
9-G1MR-25
LONGITUDINAL SECTION
500mm
9·5M
2-RT-8
G1-MR-6-500
9 RT-20
100mm
550mm
300mm
SECTION AA
2RT-20
2-RT-20
A
G1-MR-6-500
2-RT-8
650 mm
770mm
1·31 M
A
9-RT-20.
500mm
9·5M
LONGITUDINAL SECTION

14

L-Beams-General

14·1 Definition

Beams cast monolithic with slabs, to use the slab as the compression flange on one side of the rib only is called L—Beam.

Limiting stiffness, same as per T-Beams.

14·2 Flange Width

The breadth of the flange assumed as taking compression shall not execed the least of the following

(*i*) $\frac{1}{6}$ of the effective span of the L—beam

(*ii*) The breadth of the rib plus $\frac{1}{2}$ of the clear distance between ribs.

(*iii*) The breadth of the rib plus four times the thickness of the slab.

15

L–Beam for a Class Room Floor

15·1 Data

Clear span $=7$ m

Centre to centre of supports $=7{\cdot}45$ m

Spacing of T-Beam ribs $=2{\cdot}75$ m c/c

Slab thickness $=10$ cm

Materials : Concrete M 150, Grade—1 Steel

15·2 Depth Beam

$$d=\frac{\text{span}}{15}=\frac{745}{15}=49 \text{ cm}$$

$$\text{Try } d=50 \text{ cm}$$

$$\text{Rib depth}=40 \text{ cm}$$

$$b_r=25 \text{ cm}$$

15.2·1 Effective Depth

Assuming one layer of 25 mm dia steel bars effective depth

$$=50-\left(2{\cdot}5+\frac{2{\cdot}5}{2}\right)$$

$$d_e=46{\cdot}25 \text{ cm}$$

15·2·2 Effective Span

(*i*) Clear span+effective depth $=7+0{\cdot}4625=7{\cdot}4625$ m

(*ii*) Centre to centre of supports $=7{\cdot}45$ m

$$l=7{\cdot}45 \text{ m}$$

15·3 Loads

Dead load of slab (10 cm) $=240\times\frac{2{\cdot}75}{2}=330$ kg/m

Live load (class room) $=400\times\frac{2{\cdot}75}{2}=550$ kg/m

Floor finish $=60\times\frac{2\cdot 75}{2}=82{;}5$ kg/m

Rib self weight$=\frac{40}{100}\times\frac{25}{100}\times 2400=240$ kg/m

Plaster $=50\times\frac{2\cdot 75}{2}=68\cdot 75$ kg/m

Total load $w=1271\cdot 25$ kg/m say$=1275$ kg/m

15·3·1 Bending Moment

$$BM=\frac{wl^2}{8}=\frac{1275\times 7\cdot 45^2}{8}=8800 \text{ mkg}$$

15·3·2 Shear force

$$S=\frac{wl}{2}=\frac{1275\times 7\cdot 45}{2}=4750 \text{ kg}$$

15·3·3 Main Steel

$$A_t=\frac{880000}{1400\times\cdot 9\times 46\cdot 25}=15\cdot 1 \text{ cm}^2$$

Use 3—25 mm ϕ

A_t provided $=14\cdot 73$ cm²

15·4 Effective Flange Width

Least of

(*i*) $\frac{1}{6}$th of effective span$=\frac{1}{6}\times 745=124\cdot 17$ cm

(*ii*) $b_r+(\frac{1}{2}$ clear distance between ribs$)=25+\frac{1}{2}\times 250=150$ cm

(*iii*) $b_r+4\times$thickness of slab$=25+4\times 10=65$ cm

$$B=65 \text{ cm.}$$

15·5 Stresses N.A.

$$B\times d_s\times\left(n-\frac{d_s}{2}\right)=m\times A_t(d_e-n)$$

$$65\times 10\,(n-5)=19\times 14\cdot 73\,(46\cdot 25-n)$$

$$650\,n-3250=13000-281\,n$$

$$931\,n=16250$$

$$n=17\cdot 4 \text{ cm}$$

$$c'=\frac{(n-d_s)}{n}\times c=\frac{(17\cdot 4-10)}{17\cdot 4}\times c$$

$$=\frac{7\cdot 4}{17\cdot 4}\,c=0\cdot 425\,c$$

$$\bar{y}=\frac{c+2c'}{c+c'}\times\frac{d_s}{3}=\frac{c+2\times 0\cdot 425\,c}{c+0\cdot 425\,c}\times\frac{10}{3}$$

$$=\frac{1\cdot 850}{1\cdot 425}\times\frac{10}{3}=4\cdot 3 \text{ cm}$$

Lever arm $d_e-\bar{y}=46\cdot 25-4\cdot 30=41\cdot 95$ cm

$$t=\frac{880000}{41\cdot 95\times 14\cdot 73}=1424 \text{ kg/cm}^2>1400 \text{ kg/cm}^2$$

$$c=\frac{1424}{19}\times\frac{17\cdot4}{46\cdot25-17\cdot4}=45\cdot5 \text{ kg/cm}^2 < 50 \text{ kg/cm}^2$$

Therefore section is not sufficient.

15·5·1 Revised Section

$d=50$ cm, $d_e=46\cdot25$ cm

$b_r=25$ cm

Try 4--25 mm dia bars, $A_t=19\cdot63$ cm²

$$B\times d_s\times\left(n-\frac{d_s}{2}\right)=m\times A_t(d-n)$$

$$65\times10(n-5)=19\times19\cdot63\ (46\cdot25-n)$$

$$650\ n-3250=17400\times375\ n$$

$$1025\ n=20650$$

$$n=20 \text{ cm}$$

$$c'=\left(\frac{n-d}{n}\right)c=\left(\frac{20-10}{20}\right)c=0\cdot5\ c$$

$$\bar{y}=\frac{c+2\times0\cdot5\ c}{c+0\cdot5\ c}\times\frac{10}{3}=\frac{2}{1\cdot5}\times\frac{10}{3}=4\cdot45 \text{ cm}$$

$$d_e-\bar{y}=46\cdot25-4\cdot45=41\cdot8 \text{ cm}$$

$$t=\frac{88000}{41\cdot8\times19\cdot63}=1070 \text{ kg/cm}^2 < 1400 \text{ kg/cm}^2$$

$$c=\frac{1070}{19}\times\frac{20}{26\cdot25}=43 \text{ kg/cm}^2 < 50 \text{ kg/cm}^2$$

15·5·2 Shear Stress

$$q_s=\frac{S}{b_r\times a}=\frac{4750}{25\times41\cdot8}=4\cdot15 \text{ kg/cm}^2$$

15·6 Torsional Stress

Torsional moment$=M_t$

Torsional moment is produced due to DL of slab and LL on it.

Total load—rib self weight=1275—240=1035 kg/m

Total load=1035×7·45=7710 kg

$$S=\frac{7710}{2} \text{ kg}=3855 \text{ kg}$$

$$M_t=\frac{7710}{2}\times0\cdot2=771 \text{ mkg}$$

Shear stress due to torsional moment

$$q'_s=\frac{3M_t\times t_2}{\Sigma b_1t^3_1+b_2t^3_2}=\frac{3\times771\times100\times25}{65\times10^3+40\times25^3}$$

$$=\frac{5800000}{690,000}=8\cdot4 \text{ kg/cm}^2$$

Total shear stresses$=q_s+q'_s=4\cdot15+8\cdot40=12\cdot55$ kg/cm²

This is less than 4×allowable shear stress=20 kg/cm² but greater than 5 kg/cm². Full shear reinforcement shall be provided.

15·7 Stirrup Reinforcement

For Shear

Using 15 cm spacing

$$A_{w_1}=\frac{S\times s}{t\times a}=\frac{4750\times 15}{14\times \cdot 865\times 46\cdot 25}=1\cdot 27\ \text{cm}^2$$

For Torsion

$$A_{w_2}=\frac{M_t\times s}{0\cdot 8\times t\times X_1\times Y_1}=\frac{77100\times 15}{0\cdot 8\times 1400\times 20\times 41\cdot 25}=1\cdot 25\ \text{cm}^2$$

Total stirrup reinforcement $A_w=A_{w_1}+A_{w_2}$

$$=1\cdot 27+1\cdot 25=2\cdot 52\ \text{cm}^2$$

$$\text{steel/leg of stirrup}=\frac{2\cdot 52}{2}=1\cdot 26\ \text{cm}^2$$

Use 14 mm ϕ two legged stirrups at 15 cm c/c

Steel provided $=1\cdot 54$ cm² per leg.

15·8 Longitudinal Steel

Volume of steel in legs of stirrups in one cm length provided for torsion

$$=\frac{1\cdot 25}{15}\times 46\cdot 25=3\cdot 84\ \text{cm}^2$$

Use 4—12 mm ϕ.

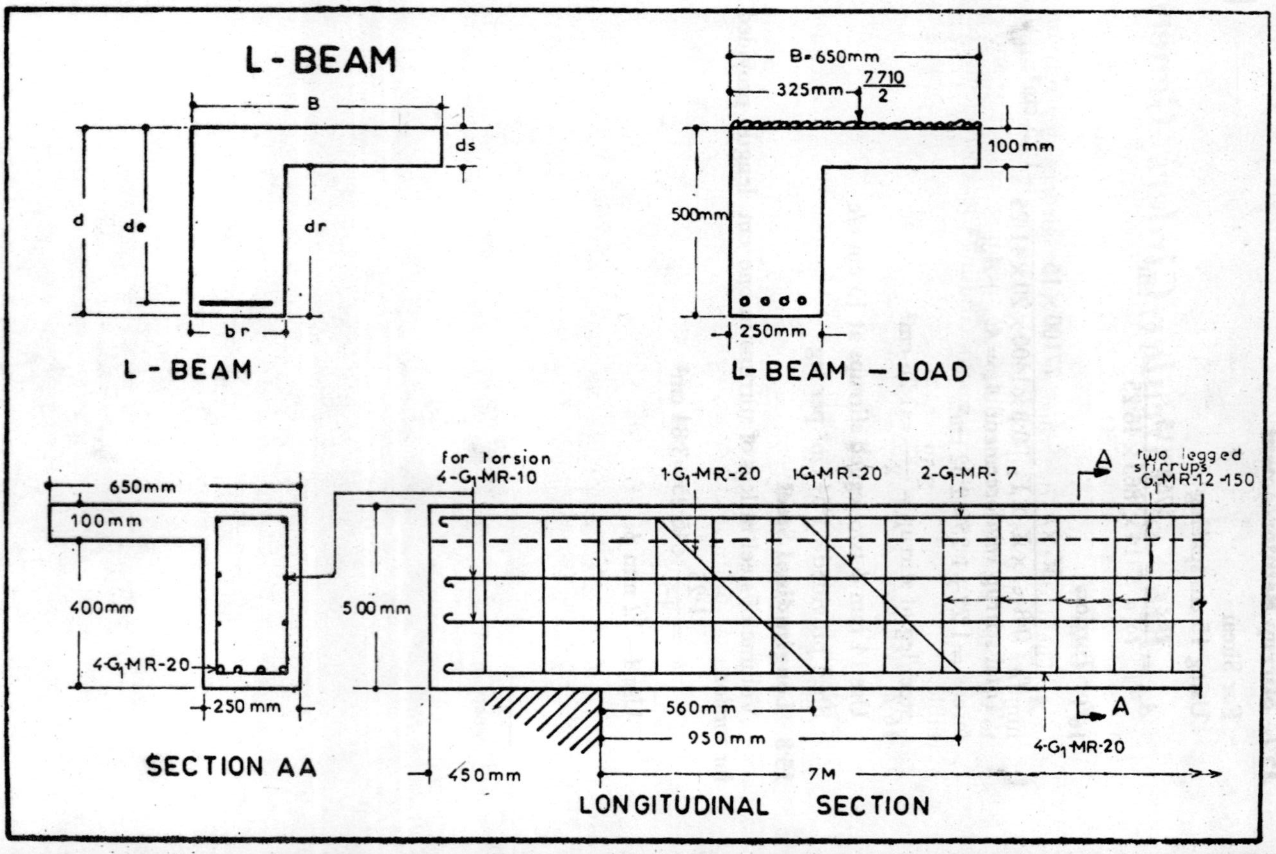

L - BEAM
B
ds
d
de
dr
br
L - BEAM
B-650mm
325mm
7710/2
100mm
500mm
250mm
L- BEAM - LOAD
650mm
100mm
400mm
4-G1-MR-20
250 mm
SECTION AA
for torsion 4-G1-MR-10
500mm
1-G1-MR-20
1-G1-MR-20
2-G1-MR- 7
A
two legged stirrups G1-MR-12-150
560mm
950mm
A
4-G1-MR-20
450mm
7M
LONGITUDINAL SECTION

16

Ring or Circular Girders—General

16·1 General

The design of the ring or circular girders depends on the number of supports on which it rests. The ring girder is subjected to torsional moments in addition to bending moments and shear forces. These moments are computed using the coefficients in the table.

16·2 Moment Coefficients

Number of supports	Bending moment at support	at centre	Torsional moment	Pt of maximum torsion
4	0·0342 WR	0·0176 WR	0·0053 WR	19°—12′
6	0·0148 WR	0·0075 WR	0·0015 WR	12°—44′
8	0·0083 WR	0·00416 WR	0·0006 WR	9°—33′
10	0·0054 WR	0·00230 WR	0·0003 WR	7°—30′
12	0·00375 WR	0·00142 WR	0·00017 WR	6°—15′

where W=total vertical load on the girder
R=Radius of the girder.

17

Ring–Beam 6·5 m Diameter

17·1 Data

Total vertical load on the beam = 178 tonnes
Diameter of the circular beam = 6·5 m
Number of supports = 6
Materials : Concrete M 200, Grade—I Steel

17·2 Relevant Codes

IS—456, IS—875

17·2·1. Allowable Stresses

$C=70$ kg/cm², $m=13$,
$t=1000$ kg/cm². $a=0{\cdot}84\ d_e$ $R=14$

17·3 Trial Section

Try 60 cm × 40 cm section

$$\text{Self weight of the beam}=\frac{60}{100}\times\frac{40}{100}\times 6{\cdot}5\times 2.4\times\pi$$
$$=11{\cdot}75 \text{ tonnes}$$

Total load on the ring beam = 178 + 11·75 = 189·75
$W=190$ tonnes (say)

$$\text{Radius of the ring beam } R=\frac{6{\cdot}5}{2}=3{\cdot}25 \text{ m}$$

17·3·1 Bending Moments

For a circular beam supported on 6 supports
Bending moment at support (−ve) = 0·0148 WR
$=\cdot 0148\times 190\times 3{\cdot}25=9{\cdot}15$ m tonnes

Bending moment at centre of supports (+ve)
$=\cdot 0075\times 190\times 3{\cdot}25=4{\cdot}62$ m tonnes

Torsional moment $=\cdot 0015\times 190\times 3{\cdot}25=0{\cdot}92$ m tonnes

17·3·2 Shear Forces

$$\text{Shear force}=\frac{\text{Total load}}{2\times\text{No. of columns}}=\frac{W}{2\times 6}=\frac{190}{12}$$
$$=15{\cdot}83 \text{ tonnes}$$

Point of maximum torsion $\theta=12{\cdot}75°$

Shear force at point of maximum torsion

$$=15{\cdot}83-\frac{15{\cdot}83\times 12{\cdot}75\times 2}{60}=15{\cdot}83-6{\cdot}75$$
$$=9{\cdot}08 \text{ tonnes}$$

17·3·3 Shear Stresses

Shear stress at point of maximum torsion

$$q=\frac{9080}{40\times 0{\cdot}87\times 56}=4{\cdot}7 \text{ kg/cm}^2$$

Shear stress due to torsional moment

$$q'=\frac{M_t\left(3+2\frac{b}{D}\right)}{b^2D}=\frac{93000\left(3+2\frac{40}{60}\right)}{40\times40\times60}$$
$$=4{\cdot}2 \text{ kg/cm}^2$$

Total shear stresses$=4{\cdot}7+4{\cdot}2$

$$=8{\cdot}9 \text{ kg/cm}^2>5 \text{ kg/cm}^2<20 \text{ kg/cm}^2$$

Therefore shear reinforcement shall be provided

$$d=\sqrt{\frac{915000}{14\times40}}=\sqrt{1633}=40{\cdot}4 \text{ cm}$$

Use $d=60$ cm (assumed value)

$d_e=54$ cm

17·3·4 Main Steel

A_t for $-$ve moment (water face)

$$=\frac{915000}{1000\times0{\cdot}84\times54}=20{\cdot}2 \text{ cm}^2$$

Use 7—20 mm ϕ at top of the support

A_t required for $+$ve moment

$$=\frac{462000}{1250\times0{\cdot}86\times54}=7{\cdot}95 \text{ cm}^2$$

Use 7 Nos. —12 mm ϕ at bottom at centre.

17·3·5 Shear Reinforcement

Shear stirrups at support. Using 12 mm ϕ 4 legged spacing

$$p=\frac{4\times1{\cdot}13\times1000\times0{\cdot}86\times54}{15830}=13.3 \text{ cm c/c}$$

Use 12 mm ϕ 4—legged at 13 cm c/c at support.

17·3·6 Shear stirrups at point of maximum Torsion

Area of steel required for shear force

$$A_{w_1}=\frac{9080\times13}{1000\times0{\cdot}86\times54}=2{\cdot}55 \text{ cm}^2$$

Using 13 cm c/c spacing area of steel required for torsional shear stress.

$$A_{w_2}=\frac{M_t\times p}{0{\cdot}8\times t_w\times X\times Y}=\frac{93000\times13}{0{\cdot}8\times1000\times50\times30}$$
$$=1{\cdot}01 \text{ cm}^2$$

Total shear reinforcement$=2{\cdot}55+1{\cdot}01=3{\cdot}56$ cm²

Using 12 mm ϕ

Area of each leg $=1{\cdot}13$ cm²

Number of legs required $=\frac{3.56}{1{\cdot}13}=3{\cdot}2$

Use 12 mm ϕ 4—legged at 13 cm c/c

7·4 Longitudinal Reinforcement for Torsion

Volume of longitudinal reinforcement per unit length provided for torsion$=\frac{1{\cdot}01\times50}{13}=3{\cdot}9$ cm²

Use 4—12 mm ϕ bars two numbers per face.

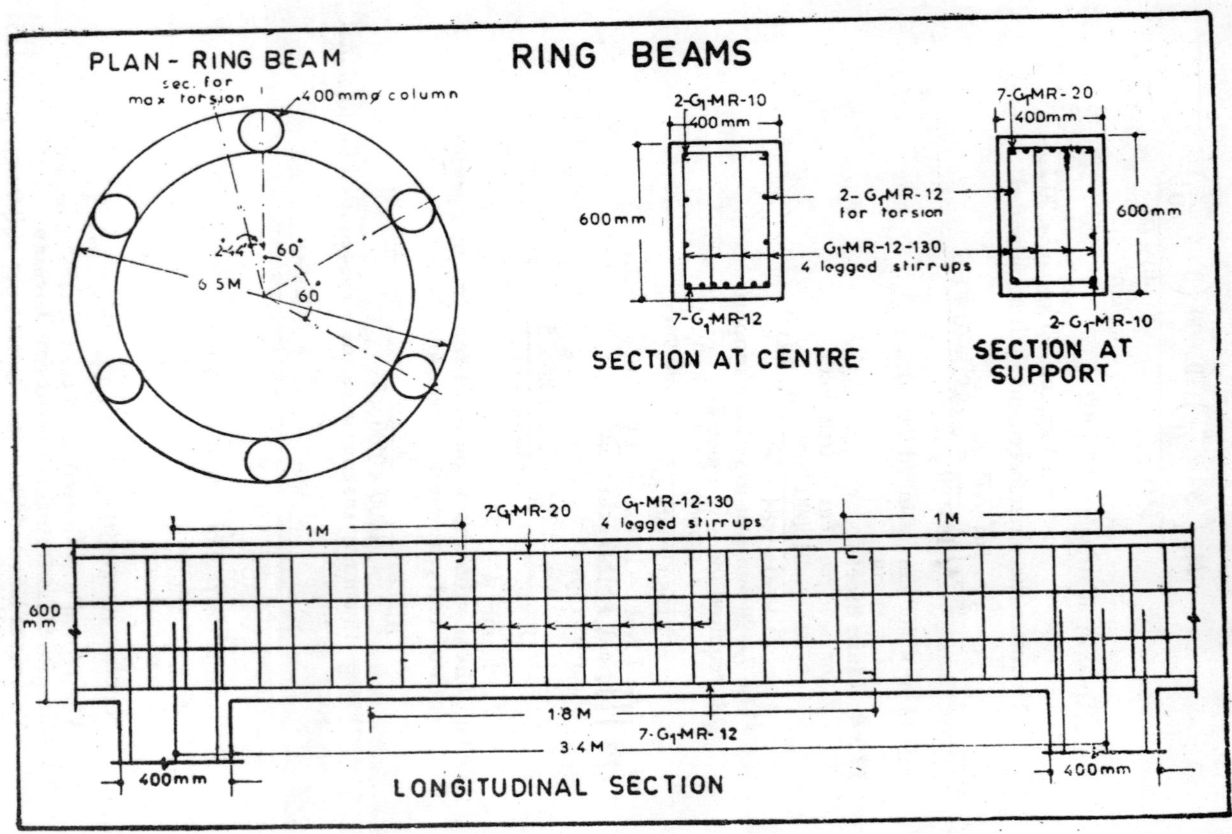
RING BEAMS
PLAN - RING BEAM
sec. for max torsion
400 mmø column
2·44
60°
60
6 5M
2-G1-MR-10
400 mm
600 mm
7-G1-MR-12
SECTION AT CENTRE
2- G1-MR-12 for torsion
G1-MR-12-130 4 legged stirrups
7-G1-MR-20
400mm
600mm
2-G1-MR-10
SECTION AT SUPPORT
1M
7-G1-MR-20
G1-MR-12-130 4 legged stirrups
1M
600 mm
1·8 M
3·4 M
7-G1-MR-12
400mm
400mm
LONGITUDINAL SECTION

III. Design of Columns and Footings

Short Columns–General

1·1 Definition

A column may be considered to be short when its effect-tive height does not exceed 15 times its least width or least lateral dimension.

1·2 Unsupported Length

(*i*) The unsupported length L of a column shall be taken as the clear distance between the slabs.

(*ii*) In beam and slab construction, it shall be the clear distance between the floor and the under side of the shallower beam framing into the column in each direction at next higher floor level.

(*iii*) In flat slab construction it shall be the clear distance between the floor and the lower extremities of the capital, the drop panel or the slab whichever is the least.

(*iv*) In columns restrained laterally by struts, it shall be the clear distance between consecutive struts in each vertical plane, provided that to be an adequate support, two such struts shall meet the column at approximately the same level and the angle, between vertical planes through the struts shall not vary more than 30° from the right angle.

Such struts shall be of adequate dimensions and shall have sufficient anchorage to restrain the column against lateral deflection.

(*v*) The columns restrained laterally by struts or beams with brackets used at the junction it shall be the clear distance between the floor and the lower edge of the bracket, provided the bracket width equals that of the beam or strut and is at least half that of the column.

1·2·1 Effective Length

The effective column length L is as in the Table

Location	Type of column	Effective Column Length L
Column of one storey	(*i*) properly restrained at both ends in position and direction.	0·75 L
	(*ii*) properly restrained at both ends in position but not in direction,	L
	(*iii*) properly restrained at one end in position and direction and imperfectly restrained in both position and direction at the other end.	A value intermediate between L and 2 L depending upon the efficiency of the imperfect restraint.

1·3 Load On Columns

(*a*) **Loads**

The loads on a building column can conveniently be divided into three following groups.

(*i*) Those due to live loads on the floor supported by the column.

(*ii*) Those due to the dead weight of the floor and beams supported by the column.

(*iii*) The weight of the columh itself.

1·3·1 Reduction in Floor Live Loads on Columns

In designing columns, the reductions in total live loads on floors may be made, as given in Table

No. of floors carried by member under consideration	% reduction of total live load on all floors above the member under consideration
1	0
2	10
3	20
4	30
5 or more	40

No reduction should be made in the case of coulmns in ware houses, garages and other buildings used for storage purposes and factories and workshops designed for a live load of 500 kg/m². However for buildings such as factories and workshops designed for a live load of more than 500 kg/m² the reductions shown as above may be made provided that the loading assumed for any column, is not less than it would have been if all floors had been designed for a live load of 500 kg/m² with no reduction.

TABLE

Location	Type of column	Effective col. length
Columns continuing through two or more stories	(*i*) properly restrained at both ends in position and direction.	0·75 *L*
	(*ii*) properly restrained at both ends in position and imperfectly restrained in direction at one or both ends.	A value intermediate between 0·75 *L* and *L* depending upon the efficiency of the directional restraint.
	(*iii*) properly restrained at one end in position and direction and imperfectly restrained in both position and direction at the other end.	A value intermediate between *L* and 2*L* depending upon the efficiency of the imperfect restraint.

1·4. Permissible Load—Axial Loads only

(*i*) Short columns with lateral ties :—The axial-load *P* permissible on a short column reinforced with longitudinal bars and lateral ties shall not exceed.

$$P=\sigma_c A_c+\sigma_{sc} A_{sc}$$

where σ_c=permissible stress in concrete in direct compression.

A_c=cross-sectional area of concrete excluding any finishing material.

σ_{sc}=permissible compressive stress for column bars =1300 kg/cm².

A_{sc}=Cross-sectional area of the longitudinal steel.

(*ii*) Short colnmns with helical reinforcement :—

$$P=\sigma_c A_k+\sigma_{sc} A_{sc}+2\sigma_{sh} A_b$$

where σ_c=permissible stress in direct compression.

A_k=Cross-sectional area of concrete in the column excluding the area of longitudinal reinforcement.

A_b=Equivalent area of helical reinforcement which may be taken as volume of helical reinforcement per unit length of the column.

A_{sc}=Cross-sectional area of steel in compression.

σ_{sh}=Permissible stress in the helical reinforcement and

σ_{sc}=Permissible stress for column bars=1300 kg/cm². The sum of the terms $\sigma_c A_k$ and $2\sigma_{sh} A_b$ should not exceed 0·5 σ_{cu} where σ_{cu} is the 28 day work cube compressive strength of concrete as in Table. All permissible stresses for concrete given in Table shall-apply.

TABLE

Grade of concrete	(15 cm cube) Cube compressive 28 day strength		Permissible stress
	Preliminary test kg/cm²	Work test	
M 100	135	100	25
M 150	250	150	40
M 200	260	200	50
M 250	320	250	60
M 300	380	300	80
M 350	440	350	90
M 400	500	400	100

1·5. Longitudinal Reinforcement

(*i*) Minimum reinforcement should not be less than ·8% of gross cross-sectional area of the column.

(*ii*) Maximum reinforcement shall not exceed 8% of the gross cross-sectional area of the column.

(*iii*) Minimum number of bars in a column having helical reinforcement, shall be at least six bars.

(*iv*) Bars less than 12 mm in diameter shall not be used.

(*v*) Column that has a larger cross-sectional area than that required to support the load, the minimum percentage of steel shall be based upon the area of concrete required to resist the direct stress and not upon the actual area.

1·6. Transverse Reinforcement

(*i*) Link or Ties

Pitch of transverse reinforcement shall not be less than the least of

(*a*) the least lateral dimension of the compression member.

(*b*) *k* times the diameter of longitudinal bar.

where

$$k=\frac{12\times\text{maximum permissible axial load on the compression member}}{\text{actual compressive load on the member (actual) axial load alone being considered in the case of member subject to combined axial load and bending)}}$$

or $k \nless 16$ and $\ngtr 24$.

(*c*) 48 times the diameter of the transverse reinforcement.

(*b*) the diameter of the link shall not be less than ¼th the diameter of longitudinal reinforcement subject to a minimum of 5 mm.

(*ii*) Helical reinforcement

(*a*) Maximum.

Pitch of the helical turns shall not be more than 7·5 cm nor greater than 1/6th the core diameter.

(*b*) **Minimum.**

Not less than 2·5 cm nor three times the diameter of steel bars forming helix.

Diameter of the helical reinforcement $\frac{1}{4}$th diameter of the longitudinal reinforcement subject to a minimum 5 mm.

1·7 Bending Moments in Columns :

These may be caused either by eccentric or bracket loads. These may also be transmitted by beams monolithic with it.

MOMENTS IN COLUMNS

	Moments for frames of one bny	Moments for frames of two or more bays
External (and similarly loaded) columns Moment at foot of upper column	$M_e \frac{K_u}{K_l+K_u+0{\cdot}5\,K_b}$	$M_e \frac{K_u}{K_l+K_u+K_l}$
Moment at head of lower columns	$M_e \frac{K_l}{K_l+K_u+0{\cdot}5\,K_b}$	$M_e \frac{K_l}{K_l+K_u+K_b}$
Internal columns :		
Moment at foot of upper column	...	$M_{es} \frac{K_u}{K_l+K_u+K_{b1}+K_{b2}}$
Moment at head of lower column.	...	$M_{es} \frac{K_l}{K_l+K_u+K_{b1}+K_{b2}}$

where M_e=bending moment at the end of the beam framing into the column assuming fixity at the connection.

M_{es}=maximum difference between the moments at the ends of two beams framing into composite sides of the column, each calculated on the assuption that the ends of the beams are fixed and assuming one of the beams unloaded.

K_u=Stiffness of the upper column

K_l=stiffness of the lower column

K_b=stiffness of the beam

K_{b1}=stiffness of tne beam on one side of the column

K_{b2}=stiffness of the beam on the other side of the column.

Bending moments in internal columns supporting on approximately symmetrical arrangement of beams and loading need not be calculated except in the case of flat slab construction.

Bending moments in external columns and in internal columns supporting an arrangement of beam and loading not approximately symmetrical should be calculated and provided for.

Columns subjected to direct load and bending should preferably be checked for ultimate load, especially when bending moments are caused predominantly by seismic load or other horizontal forces.

1·8 Design

Columns subjected to combined bending and axial loads.

(*i*) based on uncracked section

A member subjected to axial load and bending (due to eccentricity of load, monolithic construction lateral forces etc) shall be considered safe provided the following condition is satisfied,

$$\frac{\sigma_c'}{\sigma_c}+\frac{\sigma_{cb}'}{\sigma_{cb}}\leqslant 1$$

where σ_c' = calculated direct compressive stress in concrete

σ_c = premissible axial compressive stress in concrete

σ_{cb}' = calculated bending compressive stress in concrete and

σ_{cb} = permissible bending compressive stress in concrete.

The resultant tension in concrete is not greater than 35% and 25% of the resultant compression for biaxial and uniaxial bending respectively, or exceeds $\frac{3}{4}$ of the 7 day modulus of rupture of concrete.

(*ii*) based on cracked section :

If the requirements specified above are not satisfied, the stresses in concrete and steel shall be calculated by the theory of cracked section in which the tensile resistance of concrete is ignored. If the calculated stresses are within the permissible stresses specified the section may be assumed to be safe.

2

Footings—General

2·1 Definition

The foundation of a structure is the means whereby the weight of the structure and the loads upon it are transmitted to the ground. The reaction exerted by the ground must be equal and opposite to the forces exerted by the structure. To ensure even distribution of the load and to obviate unequal settlement, the centre of gravity of the loads must coincide with the centre of gravity of the upward reaction.

2·2 Safe Bearing Pressure

The foundations must always be sufficiently large to ensure that the safe bearing pressure over the ground is not exceeded.

2·3 Depth Of Foundations

Foundations must always be taken down sufficiently deep.

Rankine developed the following theory for obtaining the required depth for foundations.

$$h=\frac{p}{w}\left(\frac{1-\sin\phi}{1+\sin\phi}\right)^2$$

where p=unit vertical pressure under the footing

w=weight of soil per cubic meter.

2·4 Proportioning of Footing

Footings shall be proportioned to sustain the applied loads and induced reactions without exceeding the permissible stresses and permissible soil pressure and to ensure that any settlement which may occur shall be as nearly uniform as possible.

2·5 Footing Concentrically Loaded

In cases where the footing is concentrically loaded and the member being supported does not transmit any moment to the footing, computations for the moments and shears shall be based on an upward reaction assumed to be uniformly distributed per unit area or per pile and a downward applied load assumed to be uniformly distributed over the area of the footing covered by the column, pedestal wall or metallic column base.

2·6 Transfer of Load at the Base of Column

The compression stress in concrete at the base of column or pedestal shall be considered as being transferred by bearing to the top of the supporting pedestal or footing : The bearing pressure on the loaded area shall not exceed the permissible stress in direct compression multiplied by $\sqrt[3]{A/A'}$

(where A′=loaded area at the column base.
A=supporting area for bearing.)

of footing, which in sloped or stepped footing may be taken as the top horizontal surface of footing or assumed as the area of the lower base of the largest frustrum of a pyramid or cone contained wholly within the footing having for its upper base the area actually loaded and having side slope of one vertical to two horizontal whichever is smaller.

2·7 Wall Footings

Footings to walls are usually continuous throughout the length of the wall and take the form of a narrow slab to distribute the load over the required area.

2·8 Thickness at the Edge of Footing

In reinforced or plain concrete footings, the thickness at the edge shall not be less than 15 cm for footing on soils.

For footing on piles the thickness above the tops of piles shall not be less than 30 cm.

In the case of plain concrete pedestals the angles between the plane passing through the bottom edge of the pedestal and the corresponding junction edge of the column with pedestal and the horizontal plane shall be governed by following consideration :

$$\tan \delta \not< 0{\cdot}9 \sqrt{\frac{100 q_0}{\sigma_{cu}}+1}$$

where δ=inclination of the plane passing through the bottom edge of the pedestal and the corresponding junction edge of column to the horizontal.

q_0=maximum bearing pressure of the soil in kg/cm²

and σ_{cu}=cube strength of concrete at 28 days in kg/cm²

2·9 Computation of Bending Moments, Shear and Bond

Computation of bending moments and shears shall be based upon a uniform distribution of reaction (per unit area of the footing or per pile) for axial loading due allowance for variation in the intensity of reaction being made for the moment that may be transmitted to the footing.

2·9·1 Bending Moment

The bending moment at any section shall be determined by passing through the section a vertical plane which extends completely across the footing and computing the moment of the forces or sum of all forces acting over the entire area of the footing on one side of the said plane.

The greatest bending moment to be used in the design of an isolated concrete footing which supports a column, pedestal or wall, shall be the moment computed at sections as specified :

(*i*) At the face of the column, pedestal or wall for footing supporting a concrete column, pedestal or wall.

(*ii*) Half way between the middle and the edge of the wall, for footings under masonry wall.

(*iii*) Half way between the face of the column or pedestal and the edge of the metallic base for footings under metallic bases.

2·10 Shear

Computations for shear shall be based on a vertical section which extends completely across the footing and shear shall be taken as the sum of all the forces acting over the entire area of the footing on one side of such section.

In computing the external shear on any section through a footing supported on piles, the entire reaction from any pile whose centre is located 150 mm or more outside the section shall be assumed as producing shear on the section. The reaction from any pile whose centre is located 150 mm or more inside the section shall be assumed as producing no shear on the section. For intermediate positions of the pile centre, the portion of the pile reaction to be assumed as producing shear on the section shall be based on straight line interpolation between full value at 150 mm outside the section and zero value at 150 mm inside the section.

The critical section for shear as a measure of diagonal tension shall be assumed as a vertical section located from the face of the column, pedestal or wall at a distance equal to the depth of the footing in case of footings on soils and a distance equal to half the depth of footing for footings on piles.

3

Short RCC Column and Footing

3·1 Data

Load on the column = 50 tonnes
Shape and size = Square—30 cm × 30 cm
Bearing capacity of soil = 15 tonnes/m²
Materials : Concrete M 150, Grade—1 Steel

3·2 Relevant Codes

IS—456

Permissible Stress

$\sigma_s = 1300$ kg/cm² $\quad\quad \sigma_c = 40$ kg/cm²

3·3 Column

Column load = 50 tonnes

$P = \sigma_s A_s + \sigma_c (A_c - A_s)$

$A_c = 30 \times 30 = 900$ cm²

$50{,}000 = 1300 \times A_s + 40\ (900 - A_s)$

$$A_s = \frac{14000}{1260} = 10{\cdot}81 \text{ cm}^2$$

Use 4-12 mm ϕ ($A_s = 12{\cdot}57$ cm²)

$$\text{Minimum reinforcement} = \frac{0{\cdot}8 \times 30 \times 30}{100} = 7{\cdot}2 \text{ cm}^2$$

Steel provided is greater than minimum 0·8%
Maximum permissible load on the column.
$P = 1300 \times 12{\cdot}57 + 40\ (900 - 12{\cdot}57) = 16300 + 35500 = 51800$ kg
$= 51{\cdot}8$ tonnes > 50 tonnes.

3·3·1 Lateral Reinforcement

Least diameter of ties $= \frac{d}{4}$ where d is the diameter of the longitudinal bar or 5 mm $= \frac{20}{4} = 5$ mm

Use 5 mm ties.

3·3·2 Pitch of the Ties

Spacing of the ties shall be the least of

(i) Least lateral dimension of the column = 300 mm

(ii) k times the diameter of longitudinal bar where

$$k=\frac{12\times\text{Max. permissible load on the column}}{\text{Actual load on the column}}$$

$$=\frac{12\times 51{\cdot}8}{50}=12{\cdot}4$$

but $k=16$ minimum $=24$ maximum

Spacing $=16\times 20=320$ mm.

(iii) $48\times$ diameter of lateral reinforcement $=48\times 5$
$=240$ mm

Use 5 mm ties at 240 mm c/c.

3·4 Footing

Load on the column = 50 tonnes
Weight of footing @ 10% of column load 5 tonnes
Total load = 55 tonnes.

3·4·1 Size

$$\text{Footing area required}=\frac{\text{Load}}{\text{Bearing capacity of soil}}$$

$$=\frac{55}{15}=3{\cdot}66 \text{ sqm.}$$

$$\text{Footing side}=\sqrt{3{\cdot}66}=1{\cdot}91 \text{ m}$$

Say 2 m × 2 m

$$\text{Upward pressure of soil}=\frac{P}{A}=\frac{50}{2\times 2}=12{\cdot}5 \text{ tonnes/m}^2$$

3·4·2 BM

Maximum cantilever projection of footing from the face of the column $=\left(\frac{2-0{\cdot}3}{2}\right)=\frac{1{\cdot}7}{2}=0{\cdot}85$ m

$$\text{BM/m width}=12{\cdot}5\times{\cdot}85\times\frac{{\cdot}85}{2}=4{\cdot}5 \text{ m tonnes}$$

3·4·3 Depth Required for BM

Depth of footing at the face of column

$$d_e=\sqrt{\frac{4{\cdot}5\times 100\times 1000}{100\times 8{\cdot}74}}=\sqrt{515}=22{\cdot}6 \text{ cm}$$

Try $d=30$ cm.

3·4·4 Depth of Footing at Critical Section for SF

SF at critical section $=(2\times 2-0{\cdot}9\times 0{\cdot}9)\times 12{\cdot}5$
$=3{\cdot}19\times 12{\cdot}5=39{\cdot}7$ tonnes

$$Q=\text{SF/metre width}=\frac{39700}{4\times 0{\cdot}9}=11100 \text{ kg}$$

$$q_s=\frac{Q}{b\times{\cdot}865\, d_e}$$

$$d_e = \frac{11100}{5 \times 100 \times \cdot 865}$$
$$= 25 \cdot 8 \text{ cm}$$

Total depth required at critical section for shear
$$= 25 \cdot 8 + 7 \cdot 5 = 33 \cdot 3 \text{ cm (Say 34 cm)}$$

Total depth required at the face of the column assuming 15 cm depth at the edge of footing.

$$= 15 + (34 - 15) \times \frac{\cdot 85}{\cdot 55}$$

$$= 15 + 19 \times \frac{\cdot 85}{\cdot 55} = 15 + 29 \cdot 5 = 44 \cdot 5 \text{ cm.}$$

3·5 Shear Stress

Try $d = 40$ cm at the face of the column depth available at critical section for shear

$$d = 15 + 25 \times \frac{0 \cdot 45}{0 \cdot 85} = 15 + 13 \cdot 2 = 28 \cdot 2 \text{ cm}$$

$$d_e = 28 \cdot 2 - 7 \cdot 5 = 20 \cdot 7 \text{ cm}$$

SF at critical section $= (2 \times 2 - 1 \cdot 1 \times 1 \cdot 1)\ 12 \cdot 5$
$$= (4 - 1 \cdot 21)\ 12 \cdot 5 = 2 \cdot 79 \times 12 \cdot 5 = 34 \cdot 8 \text{ tonnes}$$

$$\text{SF/metre width} = \frac{34 \cdot 8}{1 \cdot 1 \times 4} = 7 \cdot 9 \text{ tonnes}$$

$$\text{Shear stress} = \frac{7900}{100 \times \cdot 865 \times 20 \cdot 7} = 4 \cdot 4 \text{ kg/cm}^2$$
$$< 5 \text{ kg/cm}^2.$$

Used $d = 40$ cm
$d_e = 32 \cdot 5$ cm at the face of the column
and $d = 15$ cm at the edge of the footing.

3·6 Main Stee l

$$A_t = \frac{4 \cdot 5 \times 100 \times 1000}{1400 \times \cdot 865 \times 32 \cdot 5} = 11 \cdot 4 \text{ cm}^2$$

Use 16 mm ϕ @ 17 cm c/c in both the directions.

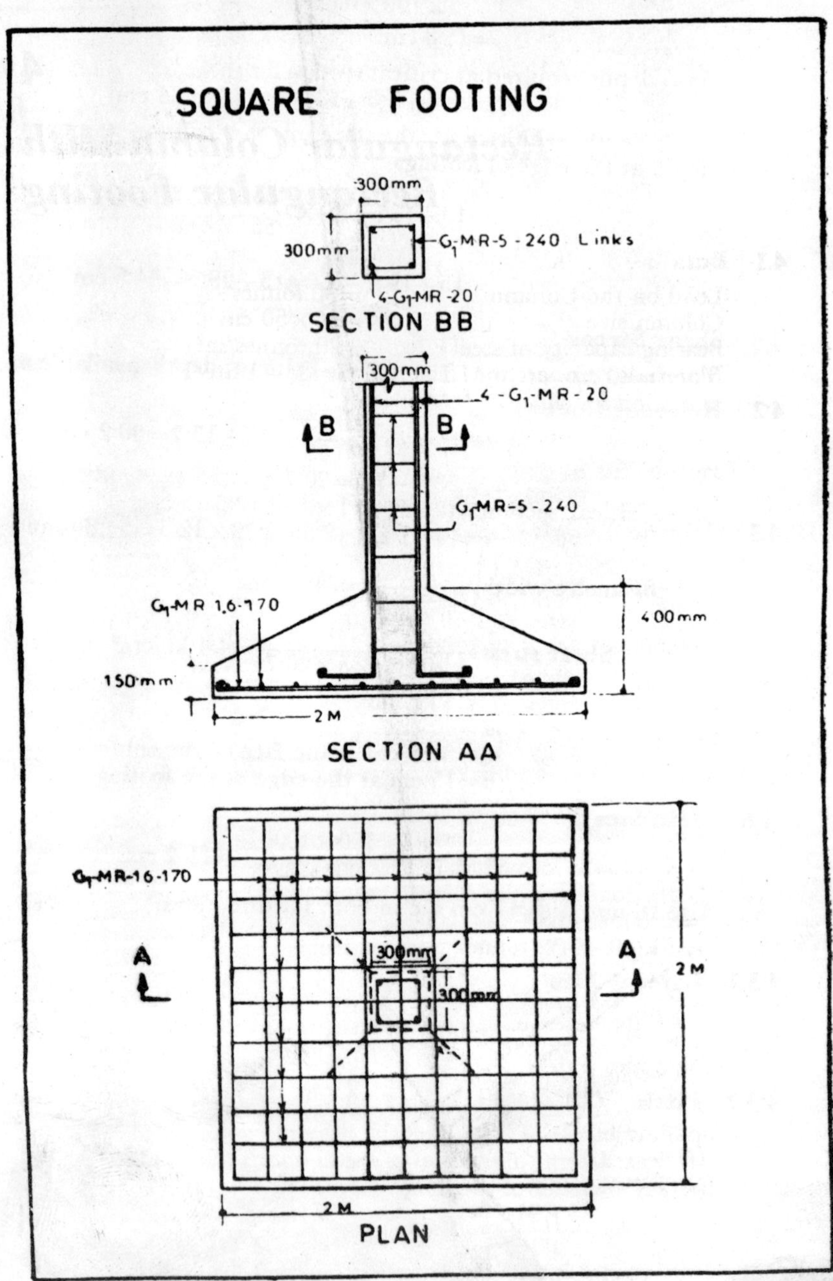
SQUARE FOOTING
300mm
300mm
G1-MR-5-240 Links
4-G1-MR-20
SECTION BB
300mm
4-G1-MR-20
B
B
G1-MR-5-240
G1-MR 16-170
400mm
150mm
2 M
SECTION AA
G1-MR-16-170
A
A
300mm
300mm
2 M
2 M
PLAN

4

Rectangular Column with Rectangular Footing

4.1 Data

Load on the Column =90 tonnes
Column size =30×50 cm
Bearing capacity of steel =20 tonnes/m²
Materials : Concrete M 150 and Grade—1 Steel

4·2 Relevant Code

IS—456

Permissible Stress

$\sigma_{sc}=1300$ kg/cm² $\sigma_c=40$ kg/cm²

4·3 Column Longitudinal steel

Load$=P=$90 tonnes

$$P=\sigma_{sc}\,A_{sc}+\sigma_c\,(A_c-A_{sc})$$
$$=1300\times A_{sc}+40\,(30\times 50-A_{sc})$$
$$90{,}000=1300\,A_{sc}+60000-40\,A_{sc}$$
$$A_{sc}=\frac{30000}{1260}=23{\cdot}8\text{ cm}^2$$

Use 4—28 mm ϕ ($A_{sc}=24{\cdot}63$ cm²)

Minimum longitudinal reinforcement in a column

$$=\frac{\cdot 8}{100}\times 30\times 50=12\text{ cm}^2$$

Steel actually provided=24·63 cm²

P=the load that the column can carry

$=1300\times 24{\cdot}63+40\,(1500-24{\cdot}63)=32200+59000$
$=91200=91{\cdot}2$ tonnes $>$ 90 tonnes

4·3·1 Lateral Ties

Least diameter of lateral ties$=\dfrac{d}{4}=\dfrac{28}{4}=7$ mm

Use 8 mm ϕ ties.

4·3.2 Pitch

Spacing of the ties shall be the least of

(*i*) least lateral dimension of the column=300 mm
(*ii*) $k\times$diameter of the longitudinal bar

$$k=\frac{12\times 91{\cdot}2}{90}=12{\cdot}2$$

but k is not less than 16
16 times the diameter$=16\times 28=448$ mm

(*ii*) 48 times the diameter of lateral ties$=48\times8=384$ mm
Use 8 mm ties at 300 mm c/c

4·4 Footing

Load$=P=90$ tonnes
Self weight of footing at 10% of column load$=9$ tonnes
Total load on the soil$=99$ tonnes

4·4·1 Size

$$\text{Footing area required}=\frac{\text{Load}}{\text{Bearing capacity of steel}}$$

$$=\frac{99}{20}=4{\cdot}95 \text{ sq m}$$

Proportioning the footing in the same proportion as the sides of the column.

$$3x\times5x=4{\cdot}95$$

$$x={\cdot}572 \text{ m}$$

Short sides of footing$=3\times{\cdot}572=1{\cdot}616$ m

Longer side of footing$=5\times{\cdot}572=2{\cdot}876$ m

Use $1{\cdot}7$ m$\times3$ m

Area$=5{\cdot}1$ m$^2>4{\cdot}95$ m^2

4·4·2 BM

$$\text{Upward soil pressure}=\frac{90}{1{\cdot}7\times3}=17{\cdot}7 \text{ tonnes/m}^2$$

Cantilever projection from the face of the short side of the Column $\left(\frac{3-{\cdot}5}{2}\right)=\frac{2{\cdot}5}{2}=1{\cdot}25$ m

BM at short side face of column ($X-X$ direction)

$$=\frac{wl^2}{2}=\frac{17700\times1{\cdot}25^2}{2}=13800 \text{ mkg.}$$

Cantilever projection from the face of the longer side of the column ($Y-Y$ direction) $=\left(\frac{1{\cdot}7-0{\cdot}3}{2}\right)=\frac{1{\cdot}4}{2}=0{\cdot}7$ m

$$\text{BM at longer side face of the column}=\frac{17700\times0{\cdot}7^2}{2}$$

$$=4320 \text{ mkg.}$$

4·4·3 Depth Required for BM

d_e=depth of footing at the face of the column

$$=\sqrt{\frac{13800\times100}{100\times8.74}}=\sqrt{1580}=39{\cdot}8 \text{ cm}$$

Try $d=50$ cm, $d_e=42{\cdot}5$ cm at face of column

4·4·4 Depth of Footing at Critical Section for Shear

SF at critical section

$$=17700\ (1{\cdot}7\times3-1{\cdot}5\times1{\cdot}3)=17700\times3{\cdot}15=56000 \text{ kg}$$

$$\text{SF/metre length of section}=\frac{56000}{(2\times1{\cdot}3+2\times1{\cdot}5)}$$

$$=\frac{56000}{5{\cdot}6}=10{,}000 \text{ kg}$$

Depth of footing required at critical section.

$$d_e=\frac{Q}{q_s\times b\times{\cdot}865}=\frac{10000}{5\times100\times{\cdot}865}$$

d required$=23+7{\cdot}5=30{\cdot}5$ cm

Try $d=50$ cm at the face of column and $d=25$ cm at the edge of footing

Depth of footing available at the critical section along shorter side.

$$d=25+25\times\frac{75}{125}=25+15=40 \text{ cm}>30{\cdot}5 \text{ cm}$$

$$d_e=40-7{\cdot}5=32{\cdot}5 \text{ cm}$$

Depth of footing available along longer side at critical section.

$$d=25+25\times\frac{20}{70}=25+7{\cdot}1=32{\cdot}1 \text{ cm}>30{\cdot}5 \text{ cm}$$

$$d_e=32{\cdot}1-7{\cdot}5=24{\cdot}6 \text{ cm}$$

4·4·5 Shear Stresses

Along short side of footing

$$q_s=\frac{10000}{100\times{\cdot}865\times32{\cdot}5}=2{\cdot}66 \text{ kg/cm}^2<5 \text{ kg/cm}^2$$

Along longer side of footing

$$q_s=\frac{10000}{100\times{\cdot}865\times24{\cdot}6}=4{\cdot}67 \text{ kg/cm}^2<5 \text{ kg/cm}^2$$

4·5 Reinforcement

Steel in the direction $Y-Y$ (longer side)

$$A_t=\frac{13800\times100}{1400\times{\cdot}865\times42{\cdot}5}=26{\cdot}8 \text{ cm}^2$$

Use 22 mm ϕ bar at 140 mm c/c

Steel in the direction $X-X$ (shorter side).

Reinforcement in the short direction in the central band equal to the width of footing (1·7 m) shall be such that

$$\frac{\text{Reinforcement in the central } 1{\cdot}7 \text{ m width}}{\text{Total reinforcement in the shorter direction}}=\frac{2}{S+1}$$

where $S=\frac{\text{Longer side of footing}}{\text{Shorter side of footing}}=\frac{3}{1\cdot7}=1\cdot76$

A_t for central 1·7 m band

$$=\text{Total steel in the shorter direction}\times\frac{2}{S+1}$$

A_t required in the shorter direction

$$=\frac{4320\times100}{1400\times\cdot865\times42\cdot5}=8\cdot25\ \text{cm}^2$$

Amount of steel to be provided in the central 1·7 m band of footing$=8\cdot25\times\frac{2}{1\cdot76+1}=6\ \text{cm}^2$

Using 12 mm ϕ bars

Number of bars$=\frac{6}{1\cdot13}=6$ Nos,

Spacing of bars$=\frac{170}{5}=34$ cm c/c

Remaining steel shall distributed in the remaining length of footing$=3-1\cdot7=1\cdot3$ m

Steel to be provided$=8\cdot25-6=2\cdot25\ \text{cm}^2$

Using 12 mm ϕ

Number of bars required$=\frac{2\cdot25}{1\cdot13}=2$

Provide 1 bar at 34 cm from the edge of the central band of 1·7 m on either side.

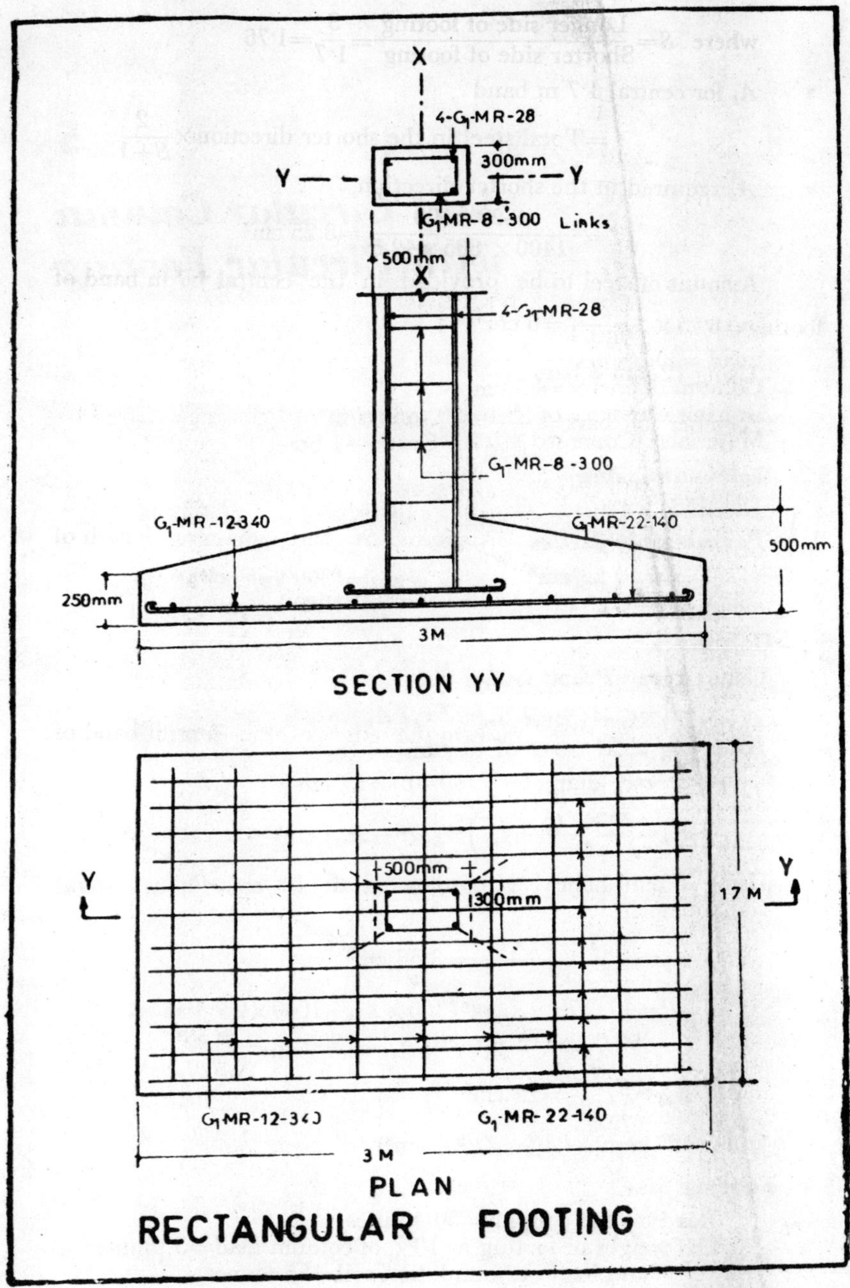
X
4-G1-MR-28
300mm
Y
Y
G1-MR-8-300 Links
500mm
X
4-G1-MR-28
G1-MR-8-300
G1-MR-12-340
G1-MR-22-140
500mm
250mm
3M
SECTION YY
500mm
300mm
Y
Y
17M
G1-MR-12-340
G1-MR-22-140
3M
PLAN
RECTANGULAR FOOTING

5

Short Circular Column with Circular Footing

5·1 Data

Load=50 tonnes
Column diameter=30 cm
Bearing Capacity of Soil=20 tonnes/m²
Materials : Concrete M 150, Grade—1 Steel

5·2 Relevant Codes

IS—456

Permissible Stress

$\sigma_c=40$ kg/cm² $\qquad \sigma_{sc}=1300$ kg/cm²
$q_s=5$ kg/cm² $\qquad \sigma_{sh}=1000$ kg/cm²

5·3 Column

Column load $P=50$ tonnes

$$P=\sigma_c\, A_k+\sigma_{sc}\, A_{sc}+2\, \sigma_{sh}\, A_b$$

Assuming clear cover of 40 mm
Concrete core diameter=(30−8)=22 cm

$$A_k=\left(\frac{\pi\times 22^2}{4}-A_{sc}\right)=(379-A_{sc})\text{ cm}^2$$

Using 8 mm helical reinforcement at 10 cm c/c as lateral reinforcement.

$$A_b=\frac{\pi(22+\cdot 8)\times 0\cdot 5}{10}=3\cdot 56\text{ cm}^2$$

$$P=40\ (379-A_{sc})+1300\times A_{sc}+1000\times 2\times 3\cdot 56$$

$$50{,}000=15100-40\ A_{sc}+1300\ A_{sc}+7120$$

$$A_{sc}=\frac{27780}{1260}=22\text{ cm}^2$$

Use 6−22 mm ϕ ($A_{sc}=22\cdot 81$ cm²)

5·4 Footing Size

Load on the column=50 tonnes
Self weight of footing at 10% of column load=5 tonnes
Total load on foundation soil=55 tonnes

$$\text{Area of circular footing}=\frac{\pi D^2}{4}=\frac{55}{20}=2\cdot 75\text{ sqm}$$

D=diameter of footing$=\sqrt{\dfrac{2{\cdot}75\times4}{\pi}}$

$=\sqrt{3{\cdot}5}=1{\cdot}87$ m

Use 2 m diameter footing.

Upward pressure of soil below the footing

$$=\frac{50000\times4}{\pi\times2^2}=15900 \text{ kg/m}^2$$

C.G. of quardrant of the footing $o\ a\ b$ from o

$$=0{\cdot}6\ \frac{(R^2+r^2+Rr)}{R+r}=0{\cdot}6\ \frac{(100^2+15^2+100\times15)}{100+15}$$

$$=\frac{0{\cdot}6\times11725}{15}=61 \text{ cm}$$

Load on the area $a'b'ab=\dfrac{\pi(1^2-0{\cdot}15^2)\times15900}{4}$

$$=\frac{\pi}{4}\times0{\cdot}9775\times15900=12200 \text{ kg.}$$

5·4·1 Bending Moment

Bending moment

$=12200\times(61-15)=561000$ cm kg

b=breadth of footing at column face (for one quadrant ab)$=a'\ b'$

$$=\frac{\pi\times D}{4}=\frac{\pi\times30}{4}=23{\cdot}5 \text{ cm}$$

$\text{BM}=8{\cdot}74\ bd^2$

$$d_e=\sqrt{\frac{561000}{8.74\times23{\cdot}5}}=\sqrt{2720}=52{\cdot}2 \text{ cm}$$

Try $d=60$ cm, $d_e=52{\cdot}5$ cm at the face of the column

and $d=20$ cm at the edge of the column

5·4·2 Depth of Slab at the Critical Section for Shear

SF at a distance 60 cm from theface of the column

$$=15900(2^2-1{\cdot}5^2)\times\frac{\pi}{4}=15900\times1{\cdot}37\times\frac{\pi}{4}=21800 \text{ kg.}$$

Shear/m width of the perimeter$=\dfrac{21800}{\pi\times1{\cdot}5}=4650$ kg

Depth required$=\dfrac{4650}{5\times{\cdot}865\times100}=10{\cdot}8$ cm

Overall depth$=10{\cdot}8+7{\cdot}5=18{\cdot}3$ cm

Depth available$=20+40\times\dfrac{25}{85}=20+11{\cdot}7=31{\cdot}7$ cm

5·4·3 Main Steel

$$A_t = \frac{561000}{1400 \times \cdot 865 \times 52 \cdot 5} = 8 \cdot 7 \text{ cm}^2$$

This amount shall be provided in both the directions in a width equal to a square inscribed in the circle of radius equal to footing radius.

Length of side of the inscribed square.

$$= R\sqrt{2} = 1 \times 1 \cdot 414 = 1 \cdot 41 \text{ m}$$

Try 12 mm ϕ

Number of bars required $= \dfrac{8 \cdot 7}{1 \cdot 13} = 8$

Spacing $= \dfrac{140}{7} = 20$ cm c/c

Use 12 mm ϕ at 200 mm c/c

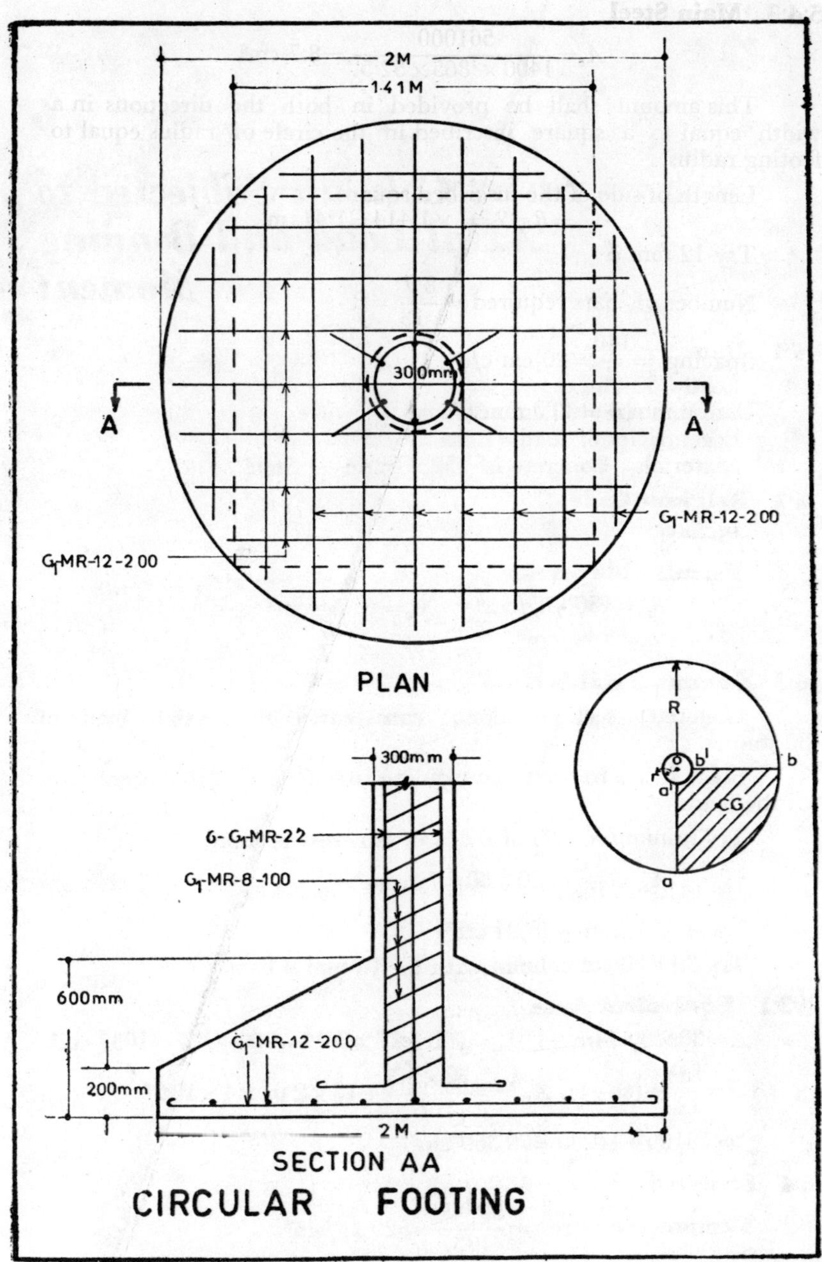

CIRCULAR FOOTING

6

Short Column Subjected to Axial Load and Bending Moment

6·1 Data

Load=30 tonnes
Column size=30 cm×30 cm
Eccentricity of load=2 cm
Materials : Concrete M 150, Grade—I Steel

6·2 Relevant Code

IS—456

Permissible Stress

$\sigma_{sc}=1300$ kg/cm² $\qquad \sigma_c=40$ kg/cm²

$q_s=5$ kg/cm²

6·3 Column Trial Section

Concrete section alone can carry the axial load of 30 tonnes.

Since it has to carry the bending moment due to eccentricity of the load.

Try minimum steel of 0·8% of concrete section.

$$A_{sc}=\frac{0{\cdot}8}{100}\times 30\times 30=7{\cdot}2 \text{ cm}^2$$

Use 4—16 mm ϕ (8·04 cm²)

Try 30×30 cm column with 4—16 mm ϕ bars.

6·3·1 Equivalent Area

$$A=30\times 30+(m-1)A_{sc}=900+17\times 8{\cdot}04=900+137=1037 \text{ cm}^2$$

$$I=\frac{bd^3}{12}+(m-1)\ A_{sc}h^2=\frac{30\times 30^3}{12}+17\times 2{\cdot}01\times 4\times 10{\cdot}2^2$$

$$=59100+14200=73{,}300 \text{ cm}^4$$

6·4 Stresses

$$\text{Compressive stress}=\frac{30{,}000}{1037}=29 \text{ kg/cm}^2$$

$$\text{Bending stress}=\frac{Pe}{z}=\frac{Pey}{I}=\frac{30000\times 2\times 15}{73300}=12{\cdot}26 \text{ kg/cm}^2$$

6·4·1 Combined Stress

$$\frac{\sigma_c'}{\sigma_c}+\frac{\sigma_b'}{\sigma_b}=\frac{29}{40}+\frac{12\cdot26}{50}=\cdot725+\cdot250=\cdot975<1$$

Use 30×30 cm column with 4—16 mm ϕ bars.

6·5 Lateral Reinforcement

$$\frac{d}{4}=\frac{16}{4}=4 \text{ mm}$$

Use 5 mm ties.

6·5·1 Pitch of the Ties

Pitch shall be the least of

(*i*) Least lateral dimension of the column=300 mm

(*ii*) 16 times the longitudinal bar=16×16=256 mm

(*iii*) 48 times the lateral reinforcement=48×5=240 mm.

Use 5 mm ϕ ties at 240 mm c/c.

7

Long Column–General

7·1 Definition

Long columns are defined as those in which the effective length of the column exceeds 15 times its least lateral dimension.

The requirement mentioned for short columns shall be used for long columns.

7·2 Permissible Stresses

The permissible stress for long columns shall be obtained by multiplying the stresses with the expression.

$C_r = 1{\cdot}5 - \frac{l}{30D}$, the permissible stresses for short columns.

Where

C_r=reduction co-efficient
l=effective length of column
D=least lateral dimension of column

If the ratio of effective column length to least radius of gyration exceeds 50, the permissible stresses shall be reduced by multiplying with the expression

$$C_r = 1{\cdot}5 - \frac{l}{100k_m}$$

where k_m=least radius of gyration.

7·2·1 Permissible Loads Axial Loads only

Permissible loads on columns shall be calculated using the expressions for short columns but using the reduced stresses, other requirements for short columns shall also apply in the design of long columns.

7·3 Concrete filled Pipe Columns

Allowable load on steel pipe filled with concrete

$$P = A_c\, \sigma_c \left(1 - 0{\cdot}000025 \frac{L^2}{K^2_e} \right) + \sigma_{sc}\, A_{sc}$$

Where A_c=area of concrete within the pipe of the column
σ_c=permissible stress in concrete in direct compression
L=unsupported length of the column as in table
K_e=radius of gyration of concrete alone

$\sigma_{sc}=\left(1200-0{\cdot}034\frac{L^2}{K^2_s}\right)$ kg/cm² provided yield strength of the pipe is not less than 2400 kg/cm² and L/K_s is not more than 120

A_{sc}=cross-sectional area of the shell of the steel pipe

K_s=radius of gyration of steel pipe alone.

7·4 Composite Columns Allowable Load

The allowable load P on a composite column consisting of a structural steel or cast iron column thoroughly encased in concrete reinforced with both longitudinal and spiral reinforcement shall not exceed that given by the following formula :

$$P=\sigma_c A_c+\sigma_{sc}\,A_{sc}+\sigma_m\,A_m$$

Where A_s=net area of concrete section

A_{sc}=cross-sectional area of longitudinal bar reinforcement.

A_m=cross-sectional area of the steel or cast iron core.

σ_m=allowable unit stress in metal core not to exceed 1250 kg/cm² for a steel core or 700 kg/cm² for a cast iron core.

σ_c=permissible stress in concrete in direct compression.

σ_{sc}=permissible compressive stress for column bars.

(*ii*) The cross-sectional area of the metal core shall not exceed 20% of gross area of the column. If a hollow metal core is used it shall be filled with concrete.

8

Long Column

8·1 Data

Load$=P=40$ tonnes
Effective weight of column $L=5$ m
Column size$=25$ cm$\times 30$ cm
Material : Concrete M 150 and Grade—1 Steel

8·2 Column

For a short column$=\frac{L}{D}=\frac{\text{Length}}{\text{Least lateral dimension}}<15$

Ratio $\frac{L}{D}=\frac{500}{25}=20>15$

Therefore it is a long column.

8·3 Permissible Stresses

Reduction co-efficient$=C_r=1{\cdot}5-\frac{L}{30D}=1{\cdot}5-\frac{500}{30\times 25}=0{\cdot}83$

$\sigma_c=0{\cdot}83\times 40=33{\cdot}2$ kg/cm^2
$\sigma_{sc}=1300\times 0{\cdot}83=1079$ kg/cm^2
Load on the column
$P=\sigma_c(A_c-A_{sc})+\sigma_{sc}A_{sc}=33{\cdot}2(750-A_{sc})+1079A_{sc}$
$40000=24900-33{\cdot}2A_{sc}+1079A_{sc}$

$A_{sc}=\frac{15100}{1045{\cdot}8}=14{\cdot}4$ cm^2

Use 4-22 mm ϕ ($A_{sc}=15{\cdot}21$ cm^2)
P=actual load that the column can carry
$=\sigma_{sc}A_{sc}+\sigma_c(A_c-A_{sc})=1079\times 15{\cdot}21+33{\cdot}2(750-15{\cdot}21)$
$=16400+734{\cdot}79\times 33{\cdot}2=16400+24500=40{,}900=40{\cdot}9$ tonnes
>40 tonnes.

8·4 Lateral Reinforcement

Least diameter of the ties$=\frac{d}{4}=\frac{22}{4}=5{\cdot}5$ mm

Use 6 mm ties.

8·4·1 Pitch

Spacing shall be the least of
(*i*) Least lateral dimension of the column$=250$ mm
(*ii*) 16 times the longitudinal bar diameter$=16\times 22=352$ mm
(*iii*) 48 times the diameter of the lateral reinforcement$=48\times 6$ $=288$ mm.

Use 6 mm ϕ at 250 mm c/c.

IV. Staircases

I

Staircases–General

1·1 Span

Staircasse flights are designed as simple slabs with a span equal to the horizontal distance between the supports.

1·2 Types

(*i*) Cantilever stairs.
(*ii*) Staircases spanning horizontally.
(*iii*) Staircases spanning longitudinally.

1·3 Width

Minimum width of a staircase=1 m

1·4 Tread and Riser

For domestic buildings
Riser=15 cm to 20 cm
Tread=22·5 cm to 30 cm

For public buildings
Riser=14 cm to 16 cm
Tread=28 cm to 32 cm

1·5 Max. Steps—Balustrade

Maximum number of steps per flight should not exceed twelve. The height of the balustrade or parapet is usually 60 cm to 90 cm.

1·6 Loads

For buildings designed for class 200 loading
Not liable for overcrowding =300 kg/m^2
Liable for overcrowding =500 kg/m^2
For all other classes of loading=500 kg/m^2

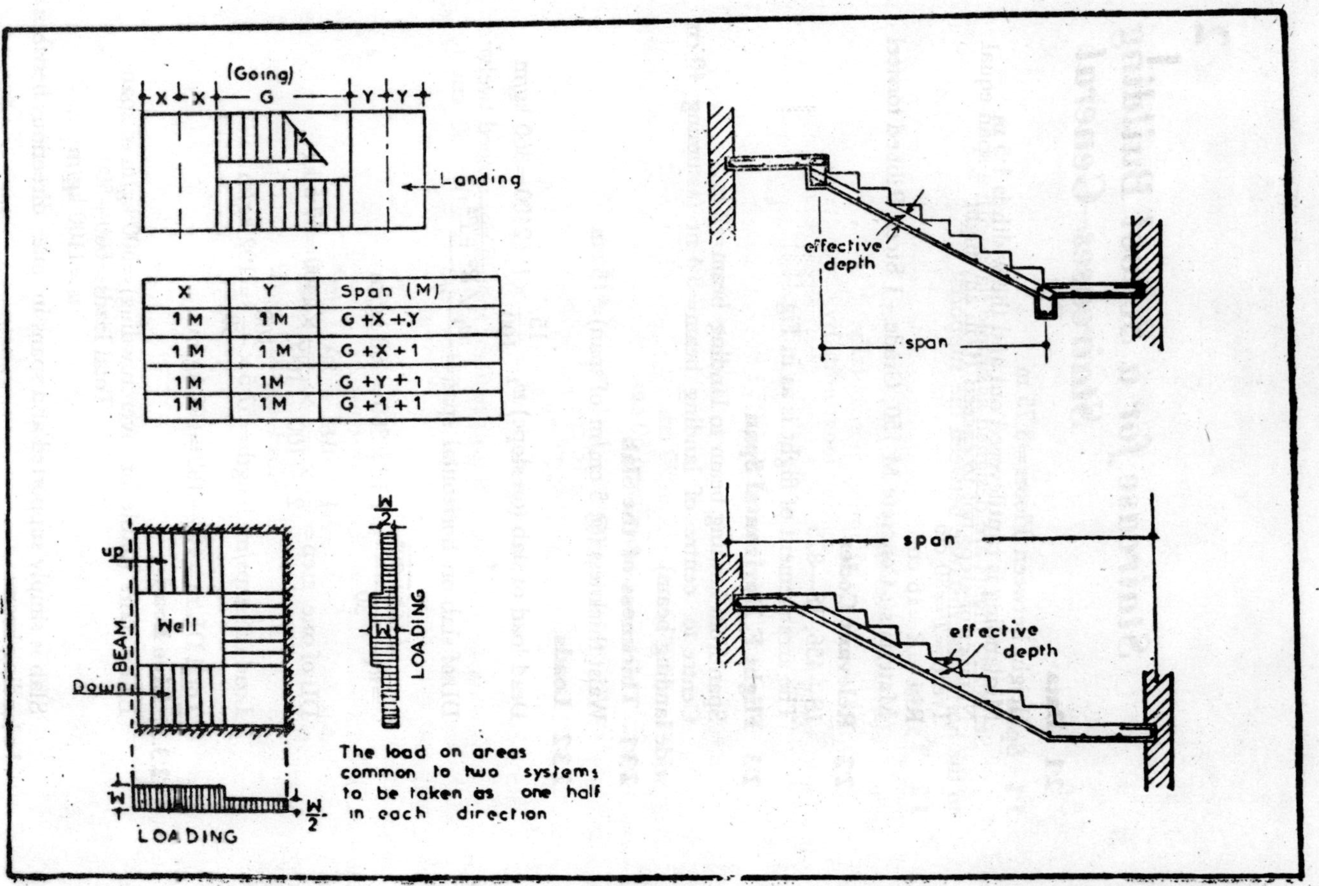
(Going)
X
X
G
Y
Y
Landing
X | Y | Span (M)
1M | 1M | G+X+Y
1M | 1M | G+X+1
1M | 1M | G+Y+1
1M | 1M | G+1+1
up
Well
BEAM
Down
W
LOADING
W/2
W
LOADING
W/2
The load on areas
common to two systems
to be taken as one half
in each direction
effective depth
span
span
effective depth

2

Staircase for a School Building

2·1 Data

Height between Floors=3·75 m
Midlanding is cantilevered out and the width is 1·5 m
Number of steps not to exceed 10 in any flight
Tread=T=30 cm
Rise=R=16 cm
Materials : Concrete M 150, Grade—1 Steel, Ribbed torsteel

2·2 Relevant Codes

IS—456, IS—875
The arrangement of flight is as in Fig.

2·3 Flight F_1 Horizontal Span

Span from landing beam to landing beam=3 m
Centre to centre of landing beam=3·4 m (assuming 40 cm wide landing beam)

2·3·1 Thickness of the Slab

Waist thickness (@ 5 cm/m of span)=15 cm

2·3·2 Loads

Dead load of slab (on slope) $w_1=\frac{15}{100}\times 1\times 2400=360$ kg/m

DL of slab on horizontal span$=\frac{w_1\sqrt{R^2+T^2}}{T}$

$$=w_1\frac{\sqrt{16^2+30^2}}{30}=1{\cdot}13\times 360=408 \text{ kg/m}$$

DL of one step$=\frac{1}{2}\times\frac{16}{100}\times\frac{30}{1000}\times 2400=57{\cdot}5$ kg

Load of steps/m length$=57{\cdot}5\times\frac{100}{30}=192$ kg/m

Total DL/m=408+192=600 kg/m

2·3.3 Live Load

LL on stairs (liable for overcrowding)=500 kg/m of span
Total loads=600+500
w=1100 kg/m

Slab is simply supported and spans in one direction between he landing beams

2·3·4 Bending Moments

$$BM=\frac{wl^2}{8}=\frac{1100\times 3{\cdot}4^2}{8}=1270 \text{ mkg}$$

2·3·5 Effective Depth

$$d_e=\sqrt{\frac{127000}{100\times 8{\cdot}74}}=\sqrt{146}=12{\cdot}1 \text{ cm}$$

Say $d_e=12{\cdot}5$ cm $d=15$ cm

2·3·6 Main Steel

$$A_t=\frac{127000}{1400\times{\cdot}865\times 12{\cdot}5}=8{\cdot}4 \text{ cm}^2$$

Use 12 mm ϕ at 13 cm c/c

2·3·7 Secondary Steel

$$A_s=\frac{0{\cdot}15}{100}\times 15\times 100=2{\cdot}25 \text{ cm}^2$$

Use 6 mm ϕ at 12·5 cm c/c

2·4 Using Torsteel

Bending moment=1270 mkg

2·4·1 Effective Depth

$$d_e=\sqrt{\frac{127000}{100\times 6{\cdot}59}}=\sqrt{93}=13{\cdot}9 \text{ cm}$$

Use $d=16$ cm $d_e=14$ cm

2·4·2 Main Steel

$$A_t=\frac{127000}{2300\times{\cdot}903\times 14}=4{\cdot}35 \text{ cm}^2$$

Use 10 mm—18 cm c/c or 12 mm—26 cm c/c

2·4·3 Secondary Steel

$$A_s=\frac{0{\cdot}15\times 16\times 100}{100}=2{\cdot}40 \text{ cm}^2$$

Use 8 mm ϕ—20 cm c/c

2·5 Flight F_2 F_3 and F_4

Centre to centre of landing beams=4·4 m
Span l=4·4 m

2·5·1 Thickness of Slab

Waist thickness (@5 cm/m of span)=4·4×5=22 cm
Use d=20 cm, d_e=18 cm

2·5·2 Loads

Dead load of slab (on slope) $w_1=\frac{20}{100}\times 1\times 2400=480$ kg/m

D.L of slab on horizontal span$=w_1\ \frac{\sqrt{R^2+T^2}}{T}$

$$=w_1 \frac{\sqrt{16^2+30^2}}{30}=1{\cdot}13\times480=542 \text{ kg/m}$$

$$\text{D.L of step}=\frac{1}{2}\times\frac{16}{100}\times\frac{30}{100}\times2400=57{\cdot}5 \text{ kg}$$

$$\text{Load of steps/m length}=57{\cdot}5\times\frac{100}{30}=192 \text{ kg/m}$$

Total DL/m=542+192=734 kg/m, say=735 kg/m

L.L=500 kg/m

Total load W=735+500=1235 kg/m

2·5·3 Bending Moment

Slab is simply supported and spans in one direction between the landing beams.

$$\text{BM}=\frac{wl^2}{8}=\frac{1235\times4{\cdot}4^2}{8}=3000 \text{ mkg.}$$

2·5·4 Effective Depth

$$d_e=\sqrt{\frac{3000\times100}{100\times8{\cdot}74}}=\sqrt{343}=18{\cdot}6 \text{ cm}$$

Use d_e=18 cm, d=20 cm

2·5·5 Main Steel

$$A_t=\frac{3000\times100}{1400\times{\cdot}865\times18}=13{\cdot}8 \text{ cm}^2$$

Use 16 mm ϕ @ 14 cm c/c

2·5·6 Secondary Steel

$$A_s=\frac{0{\cdot}15}{100}\times20\times100=3 \text{ cm}^2$$

Use 8 mm ϕ 16·5 cm c/c

2·6 Using Torsteel

BM=3000 mkg

2·6·1 Effective Depth

$$d_e=\sqrt{\frac{3000\times100}{100\times6{\cdot}59}}=\sqrt{456}=21{\cdot}5 \text{ cm}$$

Use d=24 cm, d_e=21·5 cm

2·6·2 Main Steel

$$A_t=\frac{3000\times100}{2300\times{\cdot}903\times21{\cdot}5}=6{\cdot}72 \text{ cm}^2$$

Use 12 mm ϕ 16 cm c/c

2·6·3 Secondary Steel

$$A_s=\frac{0{\cdot}15}{100}\times24\times100=3{\cdot}60 \text{ cm}^2$$

Use 8 mm ϕ 14 cm c/c

2·7 Cantilever Landing Slab S_1 Depth of Slab

Span=1·5 m

$$d=\frac{\text{span}}{10}=\frac{150}{10}=15 \text{ cm}$$

2·7·1 Loads

Live load=500 kg/m²
Dead load (15 cm)=360 kg/m²
D.L due to jali, Plaster etc. =3·225×100=322·5 kg/m²
Total=500+360+322·5=1182·5 kg/m² say 1200 kg/m²

2·7·2 Bending Moment

$$\text{B.M}=\frac{1200\times 1{\cdot}5^2}{2}=1350 \text{ kg}$$

2·7·3 Effective Depth

$$d_e=\sqrt{\frac{135000}{100\times 8{\cdot}74}}=\sqrt{155}=12{\cdot}5 \text{ cm}$$

Use $d=15$ cm $d_e=12{\cdot}5$ cm

2·7·4 Main Steel

$$A_t=\frac{135000}{1400\times{\cdot}865\times 12{\cdot}5}=8{\cdot}95 \text{ cm}^2$$

Use 12 mm ϕ at 12·5 cm c/c

2·7·5 Secondary Steel

$$A_s=\frac{0{\cdot}15}{100}\times 15\times 100=2{\cdot}25 \text{ cm}^2$$

6 mm ϕ at 12·5 cm c/c

2·8 Using Torsteel—Effective Depth

$$d_e=\sqrt{\frac{135000}{100\times 6{\cdot}59}}=\sqrt{205}=14{\cdot}3 \text{ cm}$$

Use $d_e=14{\cdot}5$ cm, $d=16{\cdot}5$ cm

2·8·1 Main Steel

$$A_t=\frac{135000}{2300\times 14{\cdot}5\times 0{\cdot}903}=4{\cdot}5 \text{ cm}^2$$

Use 12 mm—25 cm c/c or 10 mm—17 cm c/c

2·8·2 Secondary Steel

$$A_s=\frac{0{\cdot}15\times 16{\cdot}5\times 100}{100}=2{\cdot}5 \text{ cm}^2$$

Use 8 mm ϕ—20 mm c/c

2·9 Landing Beam B_1

c/c of supports=2·75 m
Clear span=2·4 m

2·9·1 Depth of Beam

$$d=\frac{\text{span}}{20}=\frac{275}{20}=13{\cdot}75 \text{ cm}$$

Due to heavy loading on the beam try $d=30$ cm, $d_e=26{\cdot}5$ cm.
Try a beam 30×40 cm

2·9·2 Effective Span

$$l=2{\cdot}4+{\cdot}265=2{\cdot}665 \text{ m}$$

2·10·1 Loads

Total dead and live loads from flights F_1 and F_2

$$=1100\times1{\cdot}2\times\frac{3{\cdot}4}{2}+1235\times1{\cdot}2\times\frac{4{\cdot}4}{2}=2260+3260=5520 \text{ kg}$$

Self weight of beam $=\frac{15}{100}\times\frac{40}{100}\times2400\times2{\cdot}665=385$ kg

Landing slab $\left(\frac{\text{span}}{10}=\frac{1{\cdot}5}{10}\right)=\frac{15}{100}\times1\times2400\times1{\cdot}5\times1{\cdot}5\times2{\cdot}75$
$=620$ kg

Dead load due to jali etc $=3{\cdot}225\times2{\cdot}75\times60=530$ kg.
Total load $=W=5520+385+620+530=7055$ kg.

2·10·1 Bending Moment

$$\text{BM}=\frac{Wl}{8}=\frac{7055\times2{\cdot}665}{8}=2360 \text{ mkg}$$

2·10·2 Effective Depth

$$d_e=\sqrt{\frac{236000}{40\times8{\cdot}74}}=\sqrt{675}=25{\cdot}9 \text{ cm}$$

Use $d_e=26{\cdot}5$ cm, $d=30$ cm

2·10·3 Main Steel

$$A_t=\frac{236000}{1400\times{\cdot}865\times26{\cdot}5}=7{\cdot}45 \text{ cm}^2$$

Use 7—12 mm ϕ

2·10·4 Shear Force

$$S=\frac{W}{2}=\frac{7055}{2}=3527{\cdot}5 \text{ kg}$$

$$q_s=\frac{3527{\cdot}5}{40\times{\cdot}865\times26{\cdot}5}=3{\cdot}85 \text{ kg/cm}^2<5 \text{ kg/cm}^2$$

Provide nominal stirrups 6 mm—20 cm c/c

2·11 Using Torsteel-Effective Depth

$$d_e=\sqrt{\frac{236000}{40\times6{\cdot}59}}=\sqrt{895}=29{\cdot}8 \text{ cm}$$

Use $d_e=30$ cm, $d=33$ cm

2·11·1 Main Steel

$$A_t=\frac{236000}{2300\times\cdot903\times30}=3\cdot47 \text{ cm}^2$$

Use 5—10 mm ϕ

2·11·2 Shear Stirrups

Provide nominal stirrups 6 mm ϕ at 20 cm c/c

2·12 Landing Beam B_2

c/c of supports=2·75 m
clear span=2·4 m

2·12·1 Depth of Beam

$$d=\frac{\text{span}}{20}=\frac{275}{20}=13\cdot75 \text{ cm}$$

Try d=30 cm, d_e=26·5 cm
Try a beam 30×40 cm

2·12·2 Effective Span

$$l=2\cdot4=0\cdot265=2\cdot665 \text{ m}$$

2·13 Loads

Total dead and live loads from flight F_2 and F_3

$$=1235\times1\cdot2\times4\cdot4=6500 \text{ kg}$$

Self weight of beam rib$=\frac{15}{100}\times\frac{40}{100}\times2400\times2\cdot665=385$ kg

Total Load=6500+385=6885 kg/m

∴ Use beam B_1

2·13·1 Beams B_3 and B_4 Loads

Effective span=2·665 m
Dead loads +live loads from flights F_2 and F_3

$$=1235\times1\cdot2\times4\cdot4=6500 \text{ kg}$$

Self wt of beam=385 kg
Landing slab=620 kg
Due to jali etc.=530 kg

Total load=8035 kg

2·13·2 Bending Moment

$$\text{BM}=\frac{Wl}{8}=\frac{8035\times2\cdot665}{8}=2680 \text{ mkg}$$

2·13·3 Effective Depth

$$d_e=\sqrt{\frac{268000}{40\times8\cdot74}}=\sqrt{765}=27\cdot8 \text{ cm}$$

Use d_e=28 cm, d=32 cm

2·13·4 Main Steel

$$A_t=\frac{268000}{1400\times\cdot865\times28}=7\cdot9 \text{ cm}^2$$

Use 12 mm—7 Nos.

2·13·5 Shear Force

$$S=\frac{W}{2}=\frac{8035}{2}=4017{\cdot}5 \text{ kg}$$

$$q_s=\frac{4017{\cdot}5}{40\times{\cdot}865\times28}=4{\cdot}17 \text{ kg/cm}^2$$

Provide nominal stirrups 6 mm—20 mm c/c

2·14 Beam B_5

Span $l=2{\cdot}665$ m

$$d=\frac{266{\cdot}5}{20}=13{\cdot}32 \text{ cm Use } d=20 \text{ cm}$$

Size=20 cm×40 cm

2·14·1 Loads

$$\text{Live load}=400\times\frac{2{\cdot}1}{2}=420 \text{ kg/m}$$

$$\text{Dead load (10 cm)}=\frac{10}{100}\times1\times2400\times\frac{2{\cdot}1}{2}=252 \text{ kg/m}$$

$$\text{Plaster etc } 60\times\frac{2{\cdot}1}{2}=63 \text{ kg/m}$$

$$\text{Self weight}=\frac{20}{100}\times\frac{40}{100}\times1\times2400=192 \text{ kg/m}$$

Total load=927 kg/m Say w=930 kg/m

2·14·2 Bending Moment

$$\text{BM}=\frac{wl^2}{8}=\frac{930\times2{\cdot}665^2}{8}=825 \text{ mkg,}$$

2·14·3 Effective Depth Needed

$$d_e=\sqrt{\frac{82500}{40\times8{\cdot}74}}=\sqrt{235}=15{\cdot}3 \text{ cm}$$

Use $d_e=16{\cdot}5$ cm, $d=20$ cm

2·14·4 Main Steel

$$A_t=\frac{82500}{1400\times{\cdot}865\times16{\cdot}5}=4{\cdot}15 \text{ cm}^2$$

6 Nos. of—10 mm ϕ

2·14·5 Shear Stress

$$S=\frac{930\times2{\cdot}665}{2}=1240 \text{ kg}$$

$$q_s=\frac{1240}{40\times{\cdot}865\times16{\cdot}5}=2{\cdot}16 \text{ kg/cm}^2$$

Provide nominal 6 mm ϕ stirrups at 15 cm c/c

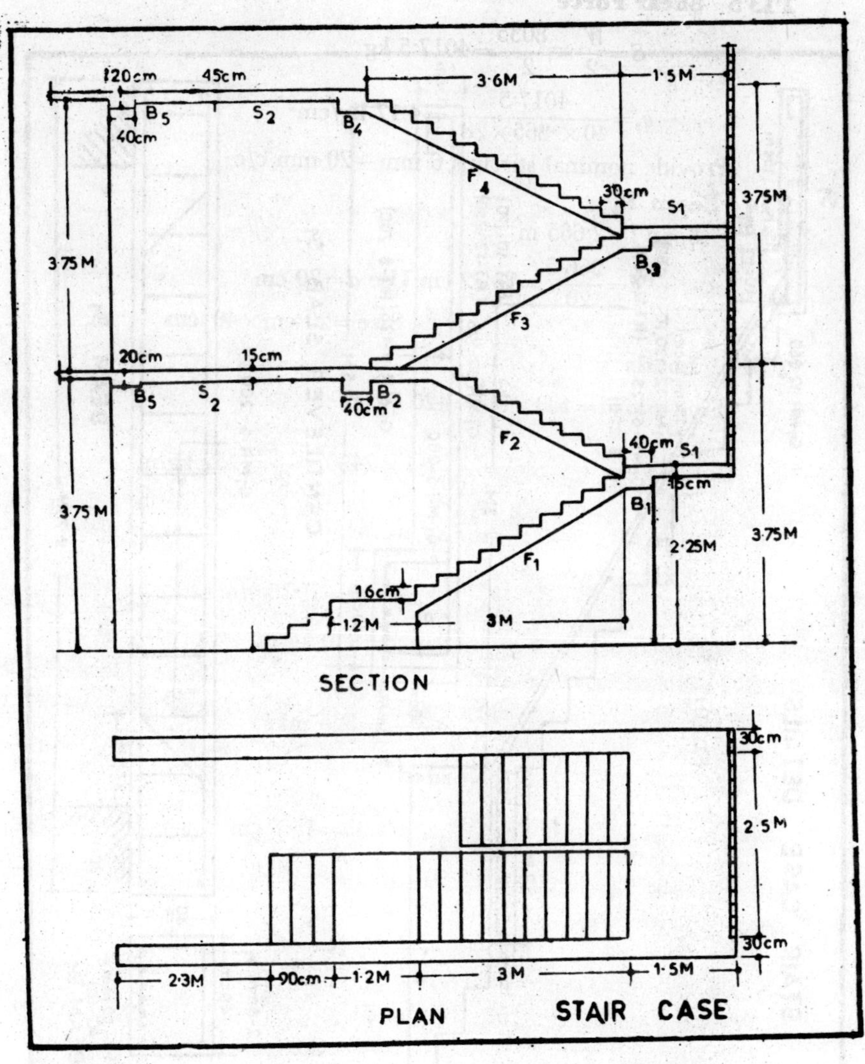

20cm
45cm
3·6M
1·5M
B_5
S_2
B_4
40cm
F_4
30cm
S_1
3·75M
3·75M
B_3
F_3
20cm
15cm
B_5
S_2
B_2
40cm
F_2
40cm
S_1
15cm
B_1
3·75M
2·25M
3·75M
F_1
16cm
1·2M
3M
SECTION
30cm
2·5M
30cm
2·3M
90cm
1·2M
3M
1·5M
PLAN
STAIR CASE

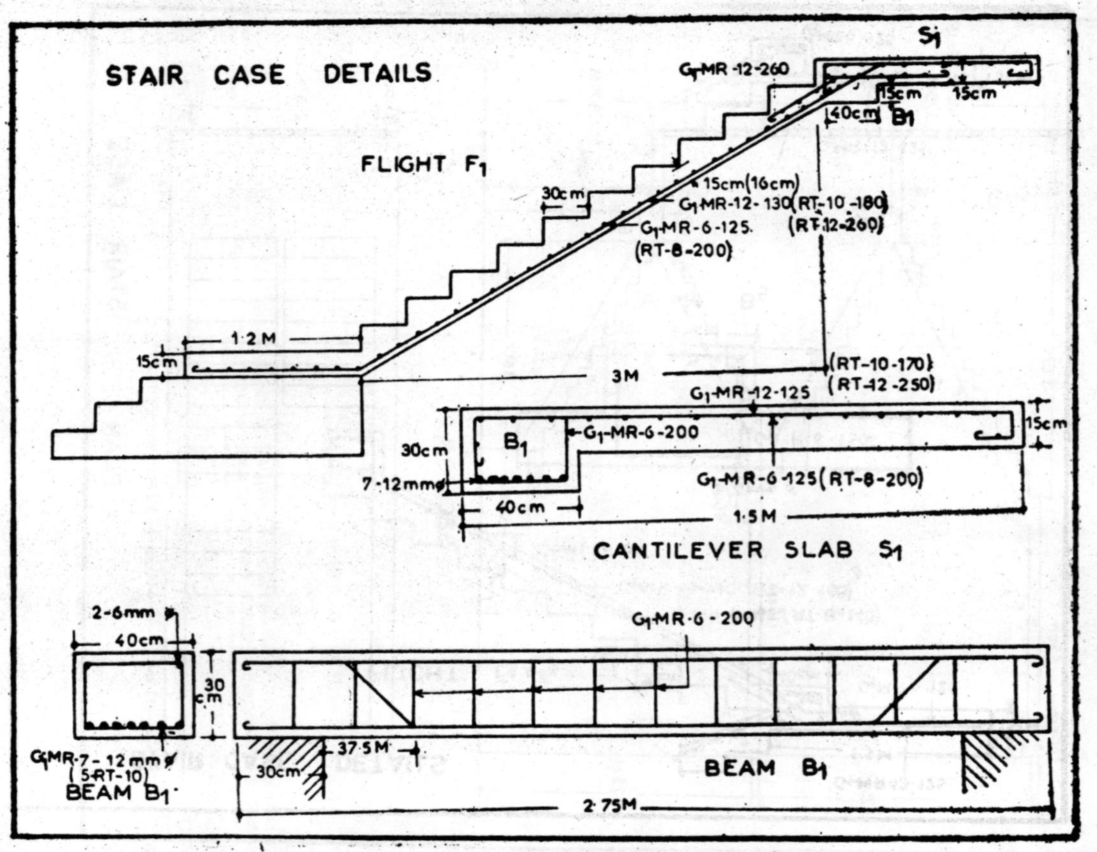

STAIR CASE DETAILS
FLIGHT F1
G1MR-12-260
S1
15cm
15cm
40cm
B1
15cm(16cm)
30cm
G1MR-12-130 (RT-10-180)
(RT-12-260)
G1MR-6-125.
(RT-8-200)
1·2 M
15cm
3M
(RT-10-170)
(RT-12-250)
G1-MR-12-125
15cm
B1
G1-MR-6-200
30cm
7-12mmφ
G1-MR-6-125 (RT-8-200)
40cm
1·5M
CANTILEVER SLAB S1
2-6mm φ
40cm
30 cm
G1MR-7-12mmφ
(5-RT-10)
BEAM B1
G1MR-6-200
37·5M
30cm
BEAM B1
2·75M

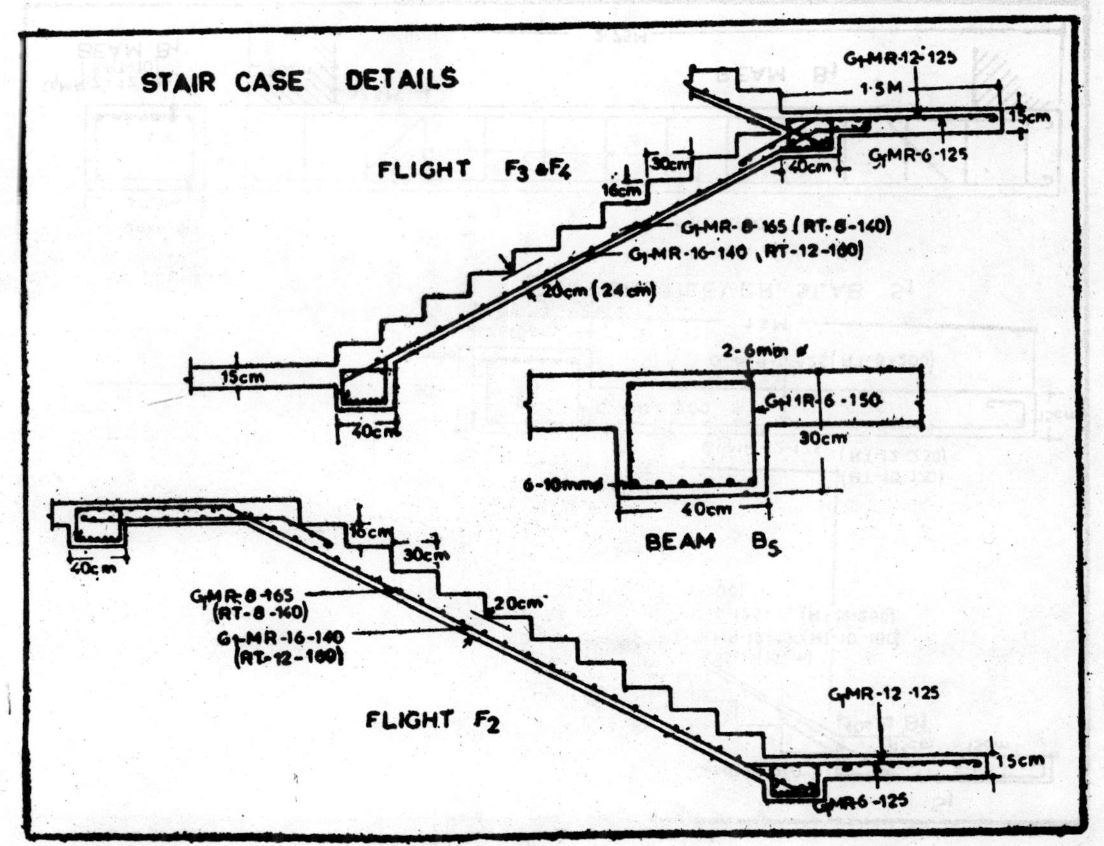
STAIR CASE DETAILS
FLIGHT F3 & F4
GMR-12-125
1·5M
15cm
GMR-6-125
40cm
30cm
16cm
GMR-8-165 (RT-8-140)
GMR-16-140 (RT-12-100)
20cm (24cm)
15cm
40cm
2-6mm ø
GMR-6-150
30cm
6-10mm ø
40cm
BEAM B5
40cm
16cm
30cm
GMR-8-165
(RT-8-140)
20cm
GMR-16-140
(RT-12-100)
FLIGHT F2
GMR-12-125
15cm
GMR-6-125

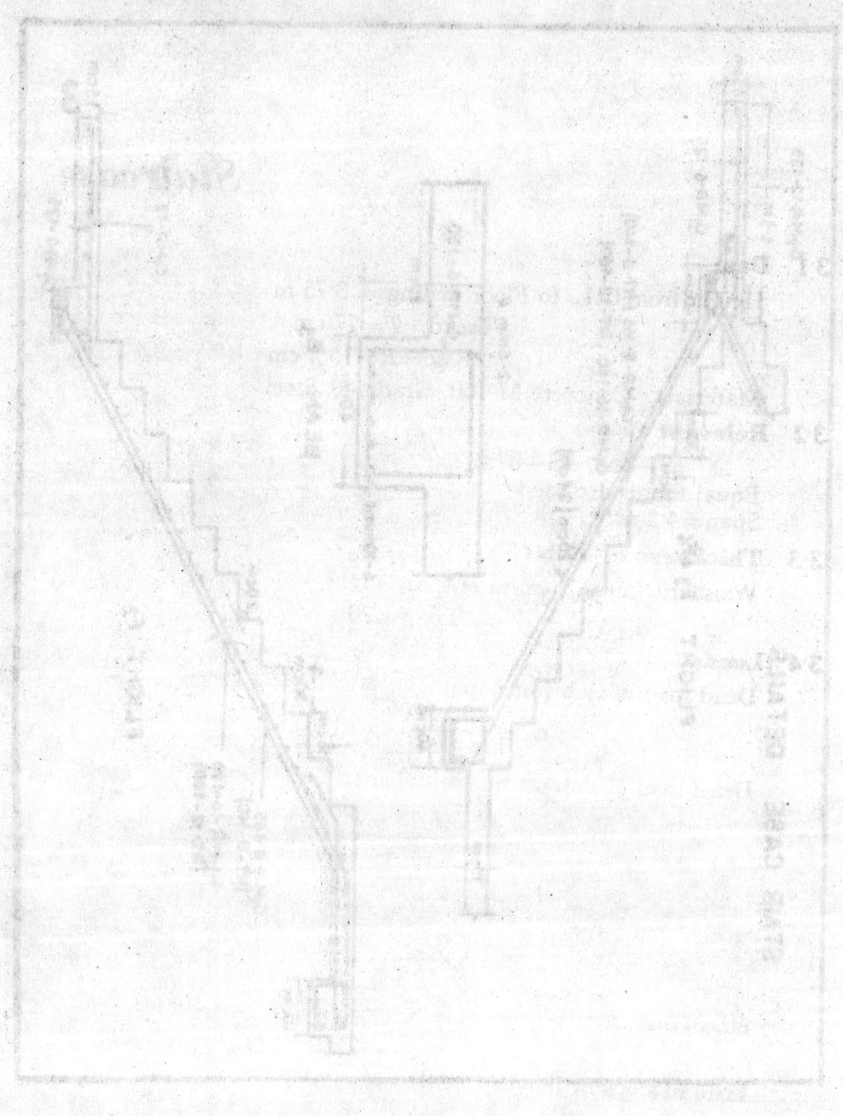

3

Staircase

3·1 Data

Height from G.L. to Floor ceiling= 3·75 m

Tread=T=27 cm

Rise=R=15·5 cm

Materials : Concrete M 150, Grade—1 Steel

3·2 Relevant Codes

IS—456, IS—875

Equal flights are used

Span=4·3 m.

3·3 Thickness of Slab

Waist thickness 5 cm/m of span=5×4·3=21·5 cm

Try d=20 cm

3·4 Loads

Dead load of slab (on slope)

$$w_1=\frac{20\times 2400}{100}=480 \text{ kg/m}$$

Dead load of slab on horizontal span

$$=w_1\frac{\sqrt{R^2+T^2}}{T}$$

$$=w_1\frac{\sqrt{15\cdot5^2+27^2}}{27}=1\cdot14\times 480$$

$$=550 \text{ kg/m}$$

$$DL \text{ of one step}=\frac{1}{2}\times\frac{15\cdot5}{100}\times\frac{27}{100}\times 2400=50\cdot2 \text{ kg/m}$$

$$\text{Load of steps/m length}=50\cdot2\times\frac{100}{27}=187 \text{ kg/m}$$

Total dead load/m=550+187 =737 kg/m

LL on stairs (liable for overcrowding)=500 kg/m of span

Total load w=737+500=1237 kg/m, Say w=1250 kg/m

3·4·1 Bending Moment

$$M=\frac{wl^2}{8}=\frac{1250\times 4\cdot3^2}{8}=2890 \text{ mkg}$$

3·4·2 Effective Depth

$$d_e = \sqrt{\frac{289000}{100 \times 8{\cdot}74}} = \sqrt{330} = 18{\cdot}2 \text{ cm}$$

Using 16 mm ϕ bars

Use $\quad d_e = 17{\cdot}9$ cm, $d = 20$ cm.

3·4·3 Main Steel

$$A_t = \frac{289000}{1400 \times {\cdot}865 \times 17{\cdot}9} = 13{\cdot}4 \text{ cm}^2$$

Use 16 mm ϕ at 15 cm c/c.

3·4·4 Secondary Steel

$$A_s = \frac{0{\cdot}15}{100} \times 20 \times 100 = 3 \text{ cm}^2$$

Use 8 mm ϕ 16·5 cm c/c.

V. Design of Structural Connections

1

Pin Connections

1·1 Introduction

A structure is an assemblage of structural members which are fastened together by some connections. Therefore the design of connections, is very important, since a designer might put forth his energy in designing very economical members to form the structure and if the joints or connections are not properly done the resulting structure will be only a poorly designed one. Therefore the designer should take utmost care in the design of the connections.

1·2 Structural Connections

In the fabrication of steel work, the commonly used types of connections are :

(*i*) Pin connections (*ii*) Bolted connections
(*iii*) Riveted connections (*iv*) Welded connections.

1·3 Pin Connections

Pins are used to fasten the structural members meeting at a point. Cylinderical shaped steel pins are used generally for those connections where relative rotation of the members meeting at a joint is permitted. In these connections it is assumed that the pin turns freely in the connection.

The diameter of pins usually vary from 3 mm to 300 mm but larger size like 600 mm dia. are not uncommon.

In general there are two types of pins. The first category of pins are used where larger rotations are allowed in the connecting member such as in members in the bascule bridges, crane booms etc.

The second type of pins are used in arched bridges. Bailey bridges, rocker supports, where lesser amount of rotation is required in connecting members.

Most of the old bridges constructed during world war including Bailey bridges used in army pins were used for connections.

1·3·1 Advantages

The advantages of pin connections are :

(*i*) Speed with which the structure can be erected
(*ii*) To make the frames statically determinate.

(*iii*) To eliminate the secondary stresses due to rigid connections due to the flexibility of the joint.

1·3·2 Disadvantages

(*i*) Due to rust formation on the pins, the pin connection does not permit free rotation.

(*ii*) Pin connections can not eliminate complete rigidity at a joint since a pin can permit rotation at the most in only one plane.

(*iii*) These connections are not desirable on account of vibrations etc. except in the case of long bridges.

1·3·3 Stresses in Pins

Pins are never subjected for direct tension. Usually they are subjected to bending.

Bending stress $$f=\frac{M_c}{I}=M\times\frac{d}{2}\Big/\frac{\pi d^4}{64}$$

$$=M\times\frac{d}{2}\times\frac{64}{\pi d^4}=\frac{10{\cdot}2\,M}{d^3}$$

The Bending Moment that the pin can take$=f\,\dfrac{d^3}{10{\cdot}2}$

where d=diameter of pin.

The shear stress in the pin is usually calculated on the basis of uniform stress distribution $f_s=\dfrac{S}{A}$

1·3·4 Bearing Stress

Assuming uniform distribution

Bearing stress $f_s=\dfrac{P}{A_b}=\dfrac{P}{t\times d}$

where P=load on the pin
t=thickness of plate
d=diameter of the pin.

1·4 Design

Size of the pin determined on the basis of bending stress provides an economical design for a pin. Using this the diameter can be obtained

$$d=\sqrt[3]{\frac{10{\cdot}2\,M}{f_b}}$$

where f_b=allowable bending stress in pins.

1·5 Problem

A pin bearing for a bridge has to be designed for a load of 300 tonnes.

1·5·1 BM

$$BM = M = 150 \times 7 = 1050 \text{ cm tonnes}$$

Bending stress, assuming the load distribution as concentrated.

$$f_b = \frac{M_y}{I} = \frac{32\,M}{\pi d^3} = 2 \text{ tonnes/cm}^2$$

if the allowable stress is 2 tonnes/cm²

$$2 = \frac{32 \times 1050}{\pi d^3}$$

$$d^3 = \frac{32 \times 1050}{\pi \times 2} = \frac{32 \times 525}{\pi}$$

or
$$d = \sqrt[3]{\frac{32 \times 525}{\pi}} = 17{\cdot}5 \text{ cm}$$

1·5·2 Shear Force

Shear force $= S = 150$ tonnes

$$f_s = \frac{S}{A} = \frac{150 \times 4}{\pi \times d^2} = \frac{600}{\pi \times d^2}$$

if $f_s = 1$ tonnes/cm², $d^2 = \dfrac{600}{\pi}$

or
$$d = \sqrt{\frac{600}{\pi}} = \sqrt{191} = 13{\cdot}8 \text{ cm}$$

1·5·3 Bearing

P = load on pin = 150 tonnes

$f_b = \dfrac{P}{d \times t'}$, Assuming bearing stress in pin

$f_b = 1{\cdot}75$ tonnes/cm²

$$1{\cdot}75 = \frac{150}{d \times 5}$$

$$d = \frac{150}{5 \times 1{\cdot}77} = 17{\cdot}1 \text{ cm.}$$

Use 18 cm pin.

2

Bolted Connections

2·1 Size

For structural steel connections the bolt sizes usually adopted are 15 mm to 25 mm in diameter. Special bolts such as anchor bolts are used for connecting steel members to the masonry parts of the structures. These bolts when used for column bases may vary in sizes from 1·5 cm to 10 cm.

2·2 Uses

Turned and fitted bolts and high strength friction slip bolts are used for avoiding slip in the connections. Bolts are used in place of rivets in such connections where rivets can not be driven and also where the rivets are subjected to tensile stresses. Ordinarily the bolts are used for connections in temporary structures. In general bolts are not widely adopted.

2·2·1 Disadvantages

(*i*) Material cost is high.
(*ii*) Reduction in strength at the root of the thread.
(*iii*) Reduction in strength due to loose fit in the holes.
(*iv*) Loosening of units due to vibration etc.

2·2·2 Advantages

(*i*) When power supply is not available for welding or riveting bolting is the only method of connection.
(*ii*) Bolted connections provide for future alterations or expansion of a structure, because of their easy removal.
(*iii*) When noise is restricted bolted connection can be advantageously adopted.

2·3 Types of Bolts

(*a*) Ordinary bolts or turned and fitted bolts.
(*b*) High strength bolts or High strength friction grip bolts.

High strength bolts provide rigid joints and initial tension can be developed in them to the strength and durability of the connection.

These are uneconomical since the material cost is almost double the cost of ordinary bolts.

2·4 Stresses

Bolted connection is subjected to bending, bearing, shearing and tension.

Bending stresses are neglected in the case of short bolts and may exist in long bolts when the grip exceeds 5 times the diameter, and should be estimated as usual.

Bearing stresses are of no significance if the loading is not excessive.

The shear is more or less uniform and is distributed over its gross area. Since the shear does not exist in the threaded portion, gross sectional area should be used in the computation of shearing stresses.

When tensile stresses are to be estimated in the bolt, the area of cross section at the root of the thread shall be taken in computations.

2·4·1 Allowable Stresses in Bolts

In Tension

Bolts over 38 mm diameter $=1260$ kg/cm²

Bolts 20 mm upto 38 mm dia $=785$ kg/cm²

In Shear

Turned and fitted bolts $=1025$ kg/cm²

Black bolts $=865$ kg/cm²

In Bearing

Turned and fitted bolts $=2360$ kg/cm²

Black bolts $=2045$ kg/cm²

When the effect of wind or seismic load is taken into account the above stresses may be exceeded by 25%.

The diameter of bolt hole shall be taken as the nominal diameter of bolt plus 1·6 mm.

Bolts in clearance holes, other than high strength friction grip bolts shall not be used for carrying a force through a connection subjected to impact or vibration or to reversal of stresses.

For purpose of distribution of load, for various types fastenings in a composite connection only rivets and turned and fitted bolts may be considered as acting together to share the load.

2·5 Design of Foundation Bolts

Holding down bolts are necessary for securely connecting the column subjected to bending moment at its base, down to the concrete block.

Oveturning moment at the column base $=M$

Stabilizing moment about the bolt line $=W\times d/2$

Net overturning moment $=M-\dfrac{Wd}{2}$

if t is the tensile load in one bolt, the moment offered by the bolts $=2\,t\times d$.

$$\therefore \quad 2t \times d = M - \frac{Wd}{2}$$

$$t = \frac{1}{2d}\left(M - \frac{Wd}{2}\right)$$

A column is subjected to moments

$$M_{xx} = 10 \text{ mt.}$$
$$M_{yy} = 6 \text{ mt.}$$

and moments are entirely to be carried by bolts.

Maximum tensile load on each bolt

$$t = \tfrac{1}{2}\left(\frac{M_{yy}}{l_y} \times \frac{M_{xx}}{l_x}\right)$$

$$t = \tfrac{1}{2}\left(\frac{6 \times 100 \times 1000}{50} + \frac{10 \times 100 \times 1000}{40}\right)$$

$$= \tfrac{1}{2}(12 + 25) = \frac{37}{2} = 18{\cdot}5 \text{ tonnes.}$$

Allowable stress in bolt in tension = 1260 kg/cm² (over 38 mm).

Area of cross-section required $= \frac{18{\cdot}5}{1{\cdot}26} = 14{\cdot}7$ sq cm.

Diameter of bolt $d = \sqrt{\frac{14{\cdot}7 \times 4}{\pi}} = \sqrt{18{\cdot}75}$

$$= 4{\cdot}35 \text{ cm} \simeq 4{\cdot}5 \text{ cm.}$$

Use 45 mm ϕ bolt.

3

Riveted Connections

·1 Introduction

Rivets are widely used in bridges and buildings. In normal steel work riveted connections are called upon to resist shear, bearing and sometimes rivets are also designed to take up tension.

3·2 Types

(*i*) Direct shear connections.

(*ii*) Moment connections or Eccentric load connections.

3·3 Assumptions

It is assumed that :

(Distribution of shearing stress across the rivet cross-section is uniform.

(*ii*) Direct load is distributed, among a group uniformly.

(*iii*) Rivet fills the hole completely.

3·4 Allowable Stresses

(*i*) **In Tension**

Axial stress on gross area of rivet,
Power driven shop rivets=785 kg/cm²
Power driven field rivets=630 kg/cm²

(*ii*) **In Shear**

Shear stress on gross area of rivets,
Power driven shop rivets=1025 kg/cm²
Power driven field rivets=945 kg/cm²
Hand driven rivets =785 kg/cm²

(*iii*) **In Bearing**

Bearing stress on gross diameter of rivets
Power driven shop rivets=2360 kg/cm²
Power driven field rivets=2125 kg/cm²

Gross and Net Areas of Rivets :

The nominal diameter of a rivet shall be taken as the diameter cold before driving.

The gross diameter of a rivet shall be taken as equal to the diameter of rivet hole.

The gross area of a rivet shall be taken as the cross-sectional area of the rivet hole.

The diameter of a rivet hole shall be taken as the nominal diameter of a rivet plus 1·5 mm for rivets of nominal diameter less than or equal to 25 mm and 2 mm for rivets of nominal diameter exceeding 25 mm unless specified otherwise.

3·4·1 Pitch and Edge Distances

Minimum Pitch

The distances between centres of rivet holes should not be less than 2·5 times the diameter of the hole.

Maximum Pitch

The distance between centres of any two adjacent rivets connecting tension or compression members shall not exceed 32 *t* or 300 mm whichever is less, where *t* is the thickness of the thinnest outside plate.

The distance between centres of two adjacent rivets in a line lying in the direction of stress shall not exceed 16 *t* or 200 mm whichever is less in tension members and 12 *t* or 200 mm, whichever is less in compression members.

Edge Distances

Dia. of hole (in mm)	Distance to sheared or hand flame cut edge from centre of hole (in mm)	Distance to rolled machine flame cut sawn or planed edge from centre of hole (in mm)
13·5 and below	19	17
15·5	25	22
17·5	29	25
19·5	32	29
21·5	32	29
23·5	38	32
25·5	44	38
29·0	51	44
32·0	57	51
35·0	57	51

3·5 Fabrication

Rivets shall be heated uniformly throughout their length, without burning or excessive sealing and shall be of sufficient length to provide a head of standard dimensions. They shall, when driven, completely fill the hole and if countersunk, the counter sinking shall be fully filled by the rivet and any projection of the counter sunk head dressed off flush, if required.

Riveted members shall have all parts firmly drawn and held together before and during riveting and special care shall be taken in this respect for all single riveted connections. For multiple riveted connections a service bolt shall be provided in every third or fourth hole.

Wherever practicable machine riveting shall be carried out by using machines of the steady pressure type.

All loose, burned or otherwise defective rivets shall be cutout and replaced before the structure is loaded and special care shall be taken to inspect all single-riveted constructions.

Special care shall be taken in heating and driving long rivets.

3·6 Selection of Rivet Size

Many factors are involved in the choice of a suitable diameter of the rivet for a given structural connection. No hard and fast rule can be laid down in the selection of rivet diameter.

It is a common practice to provide only one size of a rivet in any particular connection. Sometimes the empirical formula :

$$d=6\sqrt{t}$$

is also used in the selection of the size of the rivet

where t=thickness of the plate in mm
d=diameter of the rivet in mm

Thickness of the plate	Rivet Size	Remarks
Less than 10 mm	13 mm to 16 mm	light structures, roof trusses
10 mm to 13 mm	20 mm	Medium size structures
13 mm to 16 mm	22 mm	--do–
16 mm to 20 mm	25 mm	Heavy bridges
16 mm to 20 mm	28 mm	Very heavv bridges

3·7 Gussets

Gusset plates shall be designed to resist the shear, direct and flexural stresses acting on the weakest or critical section. Re-entrant cuts shall be avoided as far as practicable.

3·8 Lug Angles

Lug angles connecting a channel-shaped member shall, as far as possible, be disposed symmetrically with respect to the section of the member.

In the case of angle members, the lug angles and their connections to the gusset or other supporting member shall be capable of developing a strength not less than 20% in excess of the force in the outstanding lug of the angle, and the attachment of the lug angles to the angle member shall be capable of developing 40% in excess of that force.

In the case of channel members and the like, the lug angles and their connection to the gusset or other supporting member shall be capable of developing a strength of not less than 10% in excess of the force not accounted for, by the direct connection of the member and the attachment of the lug angles to the member shall be capable of developing 20% in excess of that force.

In no case shall fewer than two rivets be used for attaching the lug angle to the gusset for other supporting member.

The effective connection of the lug angle shall be, as far as possible terminated at the end of the member connected and the fastening of the lug angle to the member shall preferably start in advance of the direct connection of the member to the gusset or other supporting member.

Where lug angles are used to connect an angle member the whole area of the member shall be taken as effective.

3·9 Strength of a Riveted Joint

Strength of the joint is taken as the least of the resistances offered by it against the failures.

(*i*) Shearing of rivets

(*ii*) Tearing of rivets

(*iii*) Bearing of rivets.

(*a*) Strength in Shear

Single Shear :

When a rivet is subjected to shear along only one plane (cross-section) the rivet is said to be in single shear.

$$\text{Strength of rivet in single shear} = \frac{\pi d^2}{4} \times f_s$$

If there are n rivets in a joint then strength of a joint in single shear $= n\,\frac{\pi d^2}{4} \times f_s$

Double Shear :

When a rivet is subjected to shear across two planes (cross section) as in double covered joint, then a rivet is said to be in double shear.

$$\text{Strength of a rivet in double shear} = 2\,\frac{\pi d^2}{4} \times f_s$$

If there are n members of rivets in the joint, then strength of the joint in double shear $= n \times \frac{2\pi\, d^2}{4} \times f_s$

(*b*) Strength in tearing of a Plate

Strength of a plate against tearing $= (p-d)\, t\, f_t$

If n rivets are there in a joint then strength of joint

$$= n\,(p-d)\, t\, f_t$$

(*c*) Strength in bearing

Bearing strength of a rivet $= d \times t \times f_b$

If there are n rivets in a joint, then strength of joint in bearing

$$= n \times f_b \times d \times t.$$

4

Direct Shear Connections

4·1 Web Connection

This type of connection is a common one which is used where a simple beam frames into the web of a girder. These connections are sudjected to vertical shear due to end reactions of the beams.

4·2 Specifications

(*i*) The number of rivets required in the web of the beam is equal to the maximum end reaction or shear transmitted by beam divided by the safe-load for one rivet in bearing on the web of the beam.

These rivets are also in double shear so that shearing value of one rivet is usually much greater than the bearing value.

The rivets in the outstanding legs of the connecting angles through the web of the girder are in single shear and in bearing on the girder web or on the leg of the angle. The number of rivets required is equal to the beam shear divided by governing rivet value.

(*ii*) If there is a beam on each side of the girder web in the same location, the number of rivets required, through the girder web should be computed first by dividing the individual beam shears, by the single shear value of one rivet and then by dividing the combined beam shears by the bearing value of one rivet on the girder web ; the larger of these two number of rivets should be used.

When two beams are connected to the girder, then the rivets passing through the web of the girder must be arranged such that :

(*a*) there are enough rivets in each pair of connecting angles to satisfy the single shear requirement for the particular beam which is connected by each pair of angles.

(*b*) There are enough rivets in the entire connection to satisfy the bearing requirement for the girder web.

(*c*) The rivets are spaced so that no rivet will be too close to the lower flange of the shallower beam which would interfere with the placing and driving of that rivet.

(*iii*) The top flange of the beam must be coped, if the upper surface of the top flange of the beam is above, at or near the same elevation as the top flange of the girder.

4·3 Problem

A secondary beam ISWB—300 frames into a main beam web ISWB 600. The secondary beam transmits on end reaction of

20 tonnes. Design a suitable web connection. Use power driven shop rivets.

4·4 Data

f_s=1025 kg/cm² f_b=2300 kg/cm²
For ISWB 300, t_w=7·4 mm, ISWB 600, t_w=11·2 mm

4·5 Rivet Size

(*i*) Minimum size for structural connections=16 mm

(*ii*) $6\sqrt{t}=6\sqrt{7\cdot4}=16\cdot8$ mm

(*iii*) Plate thicknesses upto 10 mm rivet size, recommended 13 to 16 mm

Use 16 mm dia. rivets.

4·5·1 Governing Strengths of Rivet

In the web of secondary beam the rivets will be in double shear and in bearing

$$\text{Strength of rivet in double shear}=2\ \frac{\pi\times d^2}{4}\times f_s$$

$$=\frac{2\pi\times1\cdot75^2}{4}\times1025=4920\text{ kg}=4\cdot92\text{ tonnes.}$$

Strength of rivet in bearing in the web of secondary beam
$=f_b\times d\times t=2360\times1\cdot75\times0\cdot74=3\cdot05$ tonnes.

Governing strength of rivet in the secondary beam=3·05 tonnes. In the web of main beam the rivets are in single shear and in bearing.

Strength of rivet in single shear$=\pi\times1\cdot75^2\times1025$
$=2810$ kg$=2\cdot81$ tonnes.

Strength of rivet in bearing$=f\times d\times t$
$=2360\times1\cdot75\times1\cdot12=4\cdot62$ tonnes.

Governing strength of rivet in the web of main beam
$=2\cdot81$ tonnes.

4·5·2 Nnmber of Rivets Required

$$\text{Number of rivets in the secondary beam required}=\frac{20}{3\cdot05}=6\cdot6$$

Use 7—16 mm ϕ.

$$\text{Number of rivets required in the main beam}=\frac{20}{2\cdot81}=7\cdot3$$

Use 8—16 mm ϕ.

4·5·3 Length of Connecting Angles

Minimum length required to accommodate 4 rivets of 16 mm ϕ
$=2\cdot5d\times3+2\times$Edge distance
$=3\times1\cdot75\times2\cdot5+2\times2\cdot9=13\cdot11+5\cdot8=18\cdot91$ cm

Length available=300−42·9—24·95=232·15 mm=23·215 cm.
Use 20 cm length.

Width of of leg required to accommodate two rows of rivets

$=2{\cdot}5d+2\times$ Edge distance

$=2{\cdot}5\times1{\cdot}75+2\times2{\cdot}9=4{\cdot}37+5{\cdot}8=10{\cdot}17$ cm.

Width of leg required for connecting main beam to accommodate single row of rivets

$=2\times$ Edge distance $=5{\cdot}8$ cm.

Thereforc use ISA 125×75×10 mm, 20 cm long on either side of secondary beam web with 12·5 cm leg connecting it.

4·6 Problem

Two secondary beams B_1 (ISMB 400) and B_2 (ISMB 500) frame on opposite sides of the web of main beam (ISMB 600). Top flanges of all the beams are at the same level. Beam B_1 carries a reaction of 15 tonnes and beam B_2 30 tonnes. Design a suitable web connection using 22 mm diameter power driven shop rivets.

Angles of all sizes are available.

4·7 Data

ISMB 400 $t_w=8{\cdot}9$ mm, $r_1=14$ mm
ISMB 500 $t_w=10{\cdot}2$ mm, $r_1=17$ mm
ISMB 600 $t_w=12$ mm, $r_1=20$ mm

4·7·1 Allowable Stresses

Power driven shop rivets

$f_s=1025$ kg/cm², $f_b=2360$ kg/cm²

4·7·2 Relevant Codes

IS—800—1962

4·7·3 Rivet Size

Use 22 mm diameter power driven shop rivets for the entire connections.

Diameter hole $=23{\cdot}5$ mm

4·7·4 Governing Strength of Rivet

In the secondary beams the rivets are in double shear and in bearing.

Beam B_1 ISMB—400

Strength of rivet in double shear $=2\pi\dfrac{(2{\cdot}35)^2}{4}\times1025$

$=8900$ kg $=8{\cdot}9$ tonnes

Strength of rivet in bearing $=f_b\, d\, t=2360\times2{\cdot}35\times0{\cdot}89$

$=4950$ kg $=4{\cdot}95$ tonnes.

Governing strength of rivet $=4{\cdot}95$ tonnes.

Beam B_2 ISMB—500

Strength of 22 mm rivet in double shear $=8{\cdot}9$ tonnes.

Strength of rivet in bearing $=2360\times2{\cdot}35\times1{\cdot}02$

$=5660$ kg $=5{\cdot}66$ tonnes

Governing strength of rivet $=5{\cdot}66$ tonnes.

In the main beam the rivets are also in double shear and in bearing.

Beam ISMB—600

Strength in double shear=8·9 tonnes.
Strength in bearing=2360×2·35×1·2=6700 kg
=6·7 tonnes.

Governing strength of rivet =6·7 tonnes

4·8 Beam B_1 Connection

Number of rivets required to connet the beam B_1

$$=\frac{\text{Reaction}}{\text{Governing strength of rivet}}=\frac{25}{4\cdot 95}\simeq 5 \text{ Nos.}$$

Minimum length of angle required to accommodate 5 Nos. in single row=$4\times 2\cdot 5d$+edge distance×2
$=4\times 2\cdot 5\times 2\cdot 35+3\cdot 8\times 2=23\cdot 5+7\cdot 6=31\cdot 1$ cm

Minimum width of angle leg on the web=3·8×2
=7·6 cm

Length available=$40-r_1$ for ISMB 600$-r_1$ for ISMB 400
=40−2−1·4=36·6 cm

Use 36 cm long.

$$\text{Pitch}=\frac{36-7\cdot 6}{4}=\frac{28\cdot 4}{4}=7\cdot 1 \text{ cm.}$$

4·9 Beam B_2 Connection

Number of rivets required to connect the beam

$$B_2=\frac{\text{Reaction}}{\text{Governing strength of rivet}}=\frac{30}{5\cdot 66}\simeq 6$$

Minimum length of angle required to accommodate 6 rivets in single row=5×2·5×2·35+7·6=29·4+7·6=37 cm.
Length available =50—2−1·7=46·3 cm
Minimum width required=7·6 cm.
Use 45 cm long

$$\text{Pitch}=\frac{45-7\cdot 6}{5}=\frac{37\cdot 4}{5}=7\cdot 5 \text{ cm.}$$

4·10 Main Beam Connection

Total rection transferred from secondary beam to the web
=25+30=55 tonnes.

$$\text{Number of rivets required}=\frac{55}{6\cdot 7}\simeq 9$$

Use 10 Nos—5 Nos. on either side in single row

Length of angle provided will be sufficient

Use ISA 80×80×12—36 cm long on Beam B_1 side and 45 cm on B_2 side.

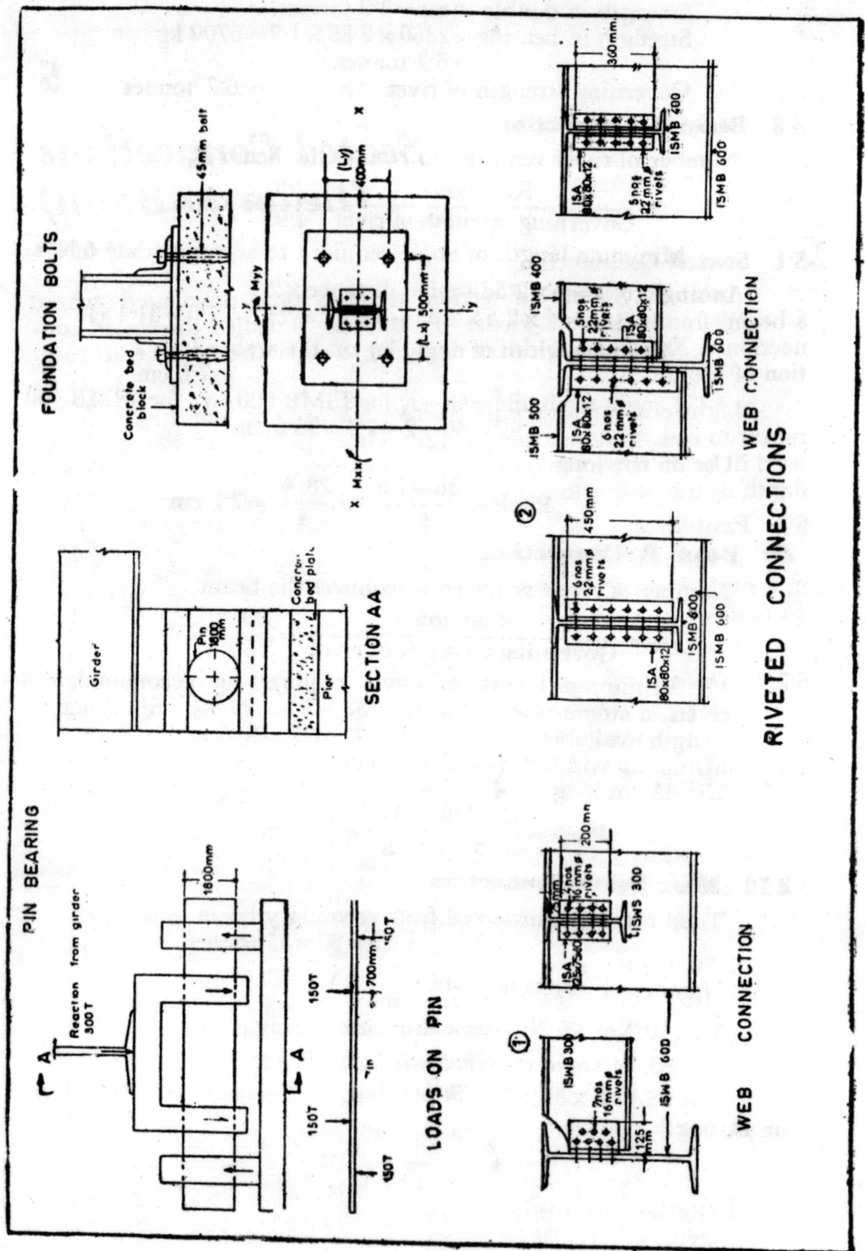
PIN BEARING
Reaction from girder 300T
1800mm
LOADS ON PIN
SECTION AA
Girder
Pier
FOUNDATION BOLTS
Concrete bed block
45mm bolt
WEB CONNECTION
RIVETED CONNECTIONS
WEB CONNECTION
ISMB 600
ISMB 400
ISMB 500

5

Seated Connections (Beam to Beam)

5·1 Seated Connection

Another type of connection which is some times used where a beam frames into a girder or column, is called the seated connection. The seat angles are designed to carry the entire end reaction of the beam.

Light clip angles are used along the web of the beam and is meant to resist lateral movement. This type of connection can be used in beam to girder connections only when there is sufficient web depth in the girder to provide for the seat angle.

5·2 Problem

A beam ISMB 400 frames into the web of a girder ISMB 600 in a steel framed building. The beam transmits a shear force of 15 tonnes on to the girder. Design a suitable connection.

Use power driven Field rivets.

5·2·1 Data

For ISMB 400, t_w=8·9 mm, b=14 cm
ISMB 600, t_w=12 mm, t_f=2·08 cm, b=21 cm

5·2·2 Allowable Stress

$$f_s = 945 \text{ kg/cm}^2$$
$$f_b = 2145 \text{ kg/cm}^2$$

5·2.3 Rivet Size

(*i*) Minimum size for structural connections =16 mm dia

(*ii*) $6\sqrt{t}$ mm $=6\sqrt{8{\cdot}9}=17{\cdot}5$ mm

(*iii*) For plate thicknesses upto 10 mm, recommended dia of rivet =13 to 16 mm.

Use 16 mm diameter rivets.

5·2·4 Governing Strength of Rivet

Strength of rivet in single shear $=\dfrac{\pi \times 1{\cdot}75^2}{4} \times 945$

$=2270$ kg $=2{\cdot}27$ tonnes.

Strength of rivet in Double shear $=2{\cdot}27 \times 2=4{\cdot}54$ tonnes

Strength of rivet in bearing (in the beam web) $=d \times t \times f_b$

$=1{\cdot}75 \times {\cdot}89 \times 2125=3320$ kg $=3{\cdot}32$ tonnes

Strength of rivet in bearing in the girder web

$=1{\cdot}75 \times 12 \times 2125=4480$ kg $=4{\cdot}48$ tonnes

5·2·5 Number of Rivets Required

Number of rivets required$=\frac{15}{2\cdot27}=6\cdot6$

Use 8—16 mm ϕ

5·3 Seat Angle

Width of flange for beam
ISMB 400=14 cm
Minimum pitch$=2\cdot5\times1\cdot75=4\cdot37$ cm
Edge distance$=2\cdot9$ cm
Minimum length of angle$=3\times4\cdot37+2\times2\cdot9=13\cdot11+5\cdot8$
$=18\cdot91$ cm
Four rivets can not be accommodated in one single row if the length of angle is equal to width of flange. Therefore provide a length=20 cm
4—16 mm ϕ rivets can easily be accommodated in one singly row. Use 2 rows.
Width of leg required$=2\cdot9+2\cdot9+4\cdot37=10\cdot17$ cm
Use ISA 110×110—12 mm thick 20 cm long.

5·3·1 Check for Thickness of Seat Angle

A gap of 12 mm shall be provided between the webs of beam and girder.
Critical section of the angle lies at the end of root fillet.
Distance from the reaction of critical section

$$=1\cdot2+\frac{b}{2}-t-r_1$$

where b=bearing length$=\frac{R}{t_wP_b}-h_2$

t=thickness of angle leg$=1\cdot2$ cm
r_1=radius at root fillet$=0\cdot8$ cm
R=reaction from the beam=15 tonnes
Beam web thickness$=t_w=0\cdot89$ cm
P_b=bearing stress$=1890$ kg/cm^2
h_2 for ISMB 400 from table$=3\cdot28$ cm

Bearing length $b=\frac{R}{t_w\,P_b}-h_2=\frac{15000}{0\cdot89\times1890}-3\cdot28$
$=8\cdot95-3\cdot28=4\cdot67$ cm

This length should not be less than

$$\frac{R}{2t_w\,P_b}=\frac{15000}{2\times0\cdot89\times1890}=4\cdot47 \text{ cm}$$

Bending moment at critical section

$$=R\left(1\cdot2+\frac{b}{2}-t-r_1\right)$$

$$=15000\left(1\cdot2+\frac{4\cdot67}{2}-1\cdot2-1\cdot0\right)$$

$$=15000(1\cdot2+2\cdot335-2\cdot2)=15000\times1\cdot335$$

$$=20000 \text{ cm kg.}$$

Moment of resistance of horizontal leg

$$=\frac{l\,t^2}{6}\times P_{bc}=\frac{20\times t^2}{6}\times 1650$$

$$=20000=\frac{20\times t^2\times 1650}{6}$$

$$t^2=\frac{20000\times 6}{20\times 1650}$$

$$t=1{\cdot}88 \text{ cm}$$

ISA 110×110×12 mm is unsuitable

Try 150×150×18 mm seat angle 14 cm long with two rows of 16 mm ϕ rivets

Bearing length $b=\frac{15000}{0{\cdot}89\times 1890}-3{\cdot}28=4{\cdot}67$ cm

Bending moment at critical section

$$=R\left(1{\cdot}2+\frac{b}{2}-1{\cdot}8-1{\cdot}2\right)$$

$$=15000\times(1{\cdot}2+2{\cdot}335-3)=15000\times{\cdot}535=8000 \text{ cm kg}$$

Moment of resistance of horizontal leg

$$=\frac{l\,t^2}{6}P_{bc}=\frac{20\times t^2}{6}\times 1650=8000$$

$$t^2=\frac{8000\times 6}{1650\times 20}=1{\cdot}46$$

$$t=1{\cdot}22 \text{ cm}<1{\cdot}8 \text{ cm used.}$$

Hence seat angle ISA 150×150×18 mm with 4–16 mm ϕ rivets will be suitable.

5·3·2 Cleat Angle

For connecting the webs of the beam and girder use ISA 100×75×8 mm, 10 cm with 2–16 mm ϕ rivets in each leg on one side or both sides.

6

Seated Connections (Columns to Beam)

6·1 Problem

Design a seated connection to support the beam ISMB 300 which transmits a load of 25 tonnes to the flange of the column ISWB 400. Use 20 mm power driven shop rivets.

6·2 Data

ISMB 300 $t_f = 12{\cdot}4$ mm, $b = 14$ cm
ISWB 400 $t_f = 13$ mm, $b = 20$ cm

6·2·1 Permissible Stresses

$$f_s = 1075 \text{ kg/cm}^2, f_b = 2360 \text{ kg/cm}^2$$

6·2·2 Governing Strength of Rivet

The rivets are in single shear. Strength of rivet in single shear$=\dfrac{\pi \times 2{\cdot}15^2}{4} \times 1075 = 3700$ kg$=3{\cdot}70$ tonnes.

Bearing strength on flange of column$= f_b\, dt = 2360 \times 2{\cdot}15 \times 13$
$= 6600$ kg$= 6{\cdot}6$ tonnes.

Governing strength of rivet$= 3{\cdot}7$ tonnes

6·2·3 Number of Rivets

Number of rivets required

$$= \frac{\text{Reaction}}{\text{Governing strength of rivet}} = \frac{25}{3{\cdot}7} \simeq 7$$

Provide the rivets in two rows

Minimum width of angle required$= 3 \times 2{\cdot}5d + 2 \times$ edge distance$= 3 \times 2{\cdot}5 \times 2{\cdot}15 + 2 \times 3{\cdot}2 = 16{\cdot}1 + 6{\cdot}4 = 22{\cdot}5$ cm

Maximum width of flange available$= 20$ cm. Therefore arrange in 3 rows

Minimum width of angle required$= 2 \times 2{\cdot}5d + 2 \times 3{\cdot}2$
$= 2 \times 2{\cdot}5 \times 2{\cdot}15 + 6\ 4 = 10{\cdot}75 + 6{\cdot}4 = 17{\cdot}15$ cm

Use 20 cm long angle

$$\text{Pitch} = \frac{20 - 6{\cdot}4}{2} = \frac{13{\cdot}6}{2} = 6{\cdot}8 \text{ cm}$$

width of leg required$= 6{\cdot}8 + 2 \times 3{\cdot}2 = 6{\cdot}8 + 6{\cdot}4 = 13{\cdot}2$ cm.

Use ISA $150 \times 150 \times 10$ mm—20 cm long

Bearing strength of angle plate$= 2360 \times 2{\cdot}15 \times 1 = 5040$ kg.
$= 5{\cdot}04$ tonnes$>$single shear strength $3{\cdot}7$ tonnes.

6·2·4 Clip angle

Provide ISA $75 \times 75 \times 6$ mm with $2-20$ mm rivets

Also use $2-20$ mm rivets to connect outstanding leg of seat angle with the flange of Beam ISMB 300.

7

Moment or Eccentric Connections

7·1 Effects of Eccentric Load

The eccentric load has two effects

(*i*) a tendency to push all the rivets in a vertical direction.

(*ii*) a turning effect, tending to turn the rivet group round the centroid of the rivets.

Sometimes a connection may pe designed to take a moment in addition to the axial load or direct force. The moment may be due to load which does not pass through the centre of gravity of the connection. Such a connection is called an eccentric connection.

7·2 Eccentric Connection

Eccentric force $=P$

Eccentricity $=e$

Moment due to eccentric force $M=P_e$

This can be replaced to act at the centre of gravity of the rivet system by

(1) Axial load P and

(2) moment $=P\times e$

Number of rivets $=n$

Distance of any rivet from the C.G. of the rivet system

Load on any rivet due to axial load P, $F_p=\dfrac{P}{n}$

Load on any rivet due to moment $P\times e=F_m$

The effect of this moment is to rotate about the C.G. of the system and to displace the rivets and the displacement is proportional to perpendicular distance r measured from the C,G. of the rivet system to the centre of rivet. The stress at any point is assumed proportional to this displacement. Since the rivets of the same diameter are generally used in a connection, the load on any rivet due to moment is also proportional to perpendicular distance.

Let F_o be the force in the rivet located at unit distance from C.G. of the system.

Force in a rivet located at distance r from C.G. of the system $=F_o\times r=F_m$.

Moment of this force about C.G. $=F_o\times r\times r=F_o\, r^2=F_m r$

Total moment of the forces in all the rivets in the system $=\Sigma F_o r^2 = F_o \Sigma r^2$

$$\therefore \quad M = F_o \, \Sigma r^2$$

$$F_o = \frac{M}{\Sigma r^2}$$

$$F_m = F_o \, r = \frac{M r}{\Sigma r^2}$$

F_m act at right angles to the radius vector, *i.e.* the line joining the C.G. of the centre of rivet.

The resultant Force in the rivet

$$F_r = \sqrt{F^2_p + F^2_m + 2F_p \, F_m \cos \theta}$$

8

Bracket Connections

8·1 Problem

A bracket transmits a load of 20 tonnes to a column ISWB 300. The load acts at a distance of 25 cm from the centre line of the column. Design the eccentric joint using power driven shop rivets.

8·2 Data

ISWB 300 $t_f=10$ mm, $b=20$ cm.

8·2·1 Permissible Stresses

$f_s=1025$ kg/m², $f_b=2360$ kg/cm²

8·2·2 Rivet Size

Use 22 mm diameter rivets

8·2·3 Governing Strength

The rivets are in single shear. Strength of rivet in single shear

$$=\frac{\pi d^2}{4} f_s=\frac{\pi\times(2{\cdot}35)^2}{4}\times 1025=4450 \text{ kg}$$

$$=4{\cdot}45 \text{ tonnes.}$$

Strength in bearing on the flange of the column

$$=f_b\, dt=2360\times 2{\cdot}35\times 1=5550=5{\cdot}55 \text{ tonnes.}$$

Governing strength of rivet=4·45 tonnes.

8·2·4 Number of Rivets

Load transmitted on each side of the flange

$$=\frac{20}{2}=10 \text{ tonnes}$$

Number of rivets required to transmit direct load

$$=\frac{10}{4{\cdot}45}\simeq 3 \text{ Nos.}$$

However some more rivets are required to take bending moment due to eccentric load.

Try 8 rivets with a symmetrical arrangement,

Minimum pitch$=2{\cdot}5d=2{\cdot}5\times 2{\cdot}35=5{\cdot}9$ cm.

Minimum edge distance=3·8 cm

Use 8 cm pitch.

and 4 cm edge distance

$$\text{Tan }\theta=\tfrac{12}{6}=2,\ \theta=63°\text{—}30',\ \cos\theta=\cdot 4462.$$

8·2·5 Resultant Load on the Extreme Rivet

Force in rivet A

$$F_a=\tfrac{10}{8}=1{\cdot}25 \text{ tonnes.}$$

distance r_a for rivet A

$$=\sqrt{12^2+6^2}=\sqrt{144+36}=\sqrt{180}=13{\cdot}4 \text{ cm}$$

distance r_b for rivet B

$$=\sqrt{6^2+4^2}=\sqrt{36+16}=\sqrt{52}=7{\cdot}2 \text{ cm}$$

$$\Sigma r^2=4\times 13{\cdot}4^2+4\times 7{\cdot}2^2=4\times 180+4\times 52=720+208$$
$$=928$$

Bending moment $M=\frac{20}{2}\times 25=250$ cm tonnes.

F_m=Force in the rivet A due to moment

$$=\frac{Mr_a}{\Sigma r^2}=\frac{250\times 13{\cdot}4}{928}=3{\cdot}61 \text{ tonnes.}$$

Resultant force in rivet A

$$F_r=\sqrt{F^2_a+F^2_m+2F_aF_m\cos\theta}$$
$$=\sqrt{1{\cdot}25^2+3{\cdot}61^2+2\times 3{\cdot}61\times 1{\cdot}25\times 0{\cdot}4462}$$
$$=\sqrt{1{\cdot}56+13{\cdot}1+4{\cdot}05}$$
$$=\sqrt{18{\cdot}71}=4{\cdot}33 \text{ tonnes.}$$

Governing strength of rivet=4·45 tonnes.

Actual load on the rivet is less than the governing strength of rivet, therefore, the arrangement is safe.

9

Crane Bracket Connections

9·1 Problem

A crane bracket is to be designed to carry a maximum load of 30 tonnes and it acts at 25 cm from the flange of column ISWB—600 supporting it. Materials available, angle sections, plates and power driven shop rivets 22 mm diameter.

9·2 Data

ISWB—600

$t_f = 23{\cdot}6$ mm, $b = 25$ cm

Plate used = 15 mm thick

Rivets = 22 mm dia.

Angles ISA 115×115×15 mm.

9·2·1 Allowable Stress

$$F_s = 1025 \text{ kg/cm}^2$$
$$f_b = 2360 \text{ kg/cm}^2.$$

9·2·2 Governing Strength of Rivets

Rivets along line *a—a* are in double shear and in bearing.

Strength in double shear

$$= 2 \times \frac{\pi \times 2{\cdot}35^2}{4} \times 1025 = 8850 \text{ kg} = 8{\cdot}85 \text{ tonnes}$$

Strength in bearing $= f_b\, dt = 2360 \times 2{\cdot}35 \times {\cdot}15 = 8350$ kg
$= 8{\cdot}35$ tonnes.

9·2·3 Rivet Line a—a

Rivets along line *a—a* have to carry a load of 30 tonnes

$$\text{Number of rivets required} = \frac{30}{8{\cdot}35} \triangleq 4.$$

9·2·4 Rivet Line b—b

These rivets are subjected to axial load and also moment due to eccentric load.

Axial load = 30 tonnes

$$\text{Moment} = 30 \times \left[25 - \frac{(11{\cdot}5 - 1{\cdot}5)}{2}\right]$$
$$= 30 \times 20 = 600 \text{ cm tonnes.}$$

Try 8 rivets along rivet line *b—b*

Minimum pitch $= 2{\cdot}5 \times 2{\cdot}35 = 5{\cdot}9$ cm.

Minimum edge distance $= 3{\cdot}8$ cm.

Load on the rivet due to axial load

$$F_a = \tfrac{30}{8} = 3{\cdot}75 \text{ tonnes.}$$

Force due to moment

$$F_m = \frac{Mr}{\Sigma r^2} = \frac{640 \times 28}{\Sigma\,(4^2+12^2+20^2+28^2)\,2}$$

$$= \frac{640 \times 28}{2684} = 6{\cdot}25 \text{ tonnes.}$$

$$F_r = \text{Resultant force} = \sqrt{6{\cdot}25^2+3{\cdot}75^2}$$

$$= \sqrt{39+14{\cdot}1} = \sqrt{53{\cdot}1} = 7{\cdot}31 \text{ tonnes.}$$

9·2·5 Rivet Line c—c

Rivets connecting vertical angle to flange of column.

Assume that all rivets are in tension, *i.e.*, N.A. passes through the bottom most rivet.

Try 20 rivets—10 rivets on either side.

Maximum force in the extreme top rivet due to moment

$$F_m = \frac{Mr}{2\Sigma r^2}$$

$$M = 30 \times 25 = 750 \text{ cm/tonnes.}$$

$$\Sigma r^2 = 8^2+16^2+24^2+32^2+40^2+48^2+56^2+64^2+72^2 = 19220$$

$$r = 72$$

$$F_m = \frac{750 \times 72}{2 \times 19220} = 1{\cdot}4 \text{ tonnes.}$$

$$F_a = \text{Force due to direct load}$$

$$= \frac{30}{20} = 1{\cdot}5 \text{ tonnes.}$$

Stress due to direct load $= \dfrac{1{\cdot}5 \times 1000 \times 4}{\pi \times (2{\cdot}35)^2} = 345 \text{ kg/cm}^2$

Stress due to moment $= \dfrac{1{\cdot}4 \times 1000 \times 4}{\pi \times (2{\cdot}35)^2} = 321 \text{ kg/cm}$

Combined stress $= \dfrac{345}{1025} + \dfrac{321}{785}$

$= 0{\cdot}336 + 0{\cdot}41 = {\cdot}746 < 1.$

Hence the arrangement is safe.

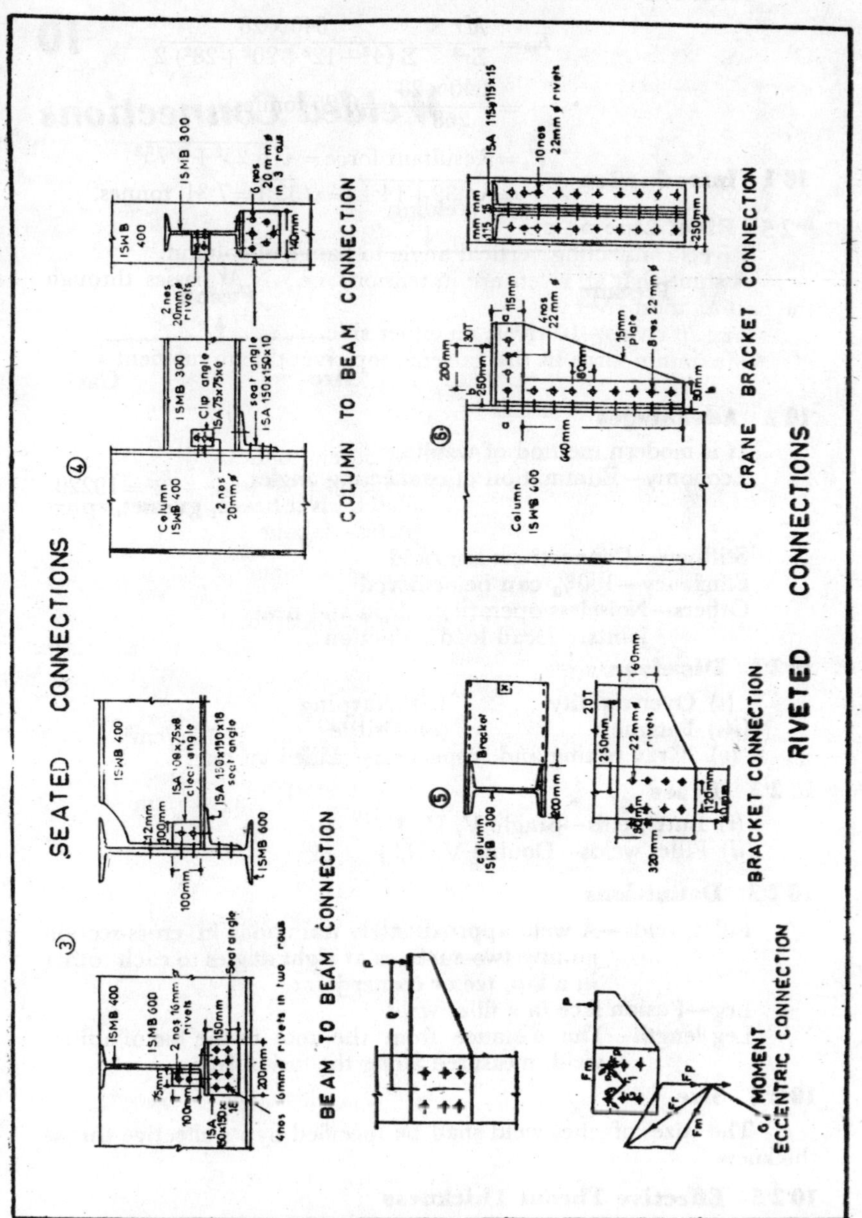
SEATED CONNECTIONS
③
BEAM TO BEAM CONNECTION
④
COLUMN TO BEAM CONNECTION
⑤
BRACKET CONNECTION
⑥
CRANE BRACKET CONNECTION
MOMENT
ECCENTRIC CONNECTION
RIVETED CONNECTIONS

10

Welded Connections

10·1 Introduction

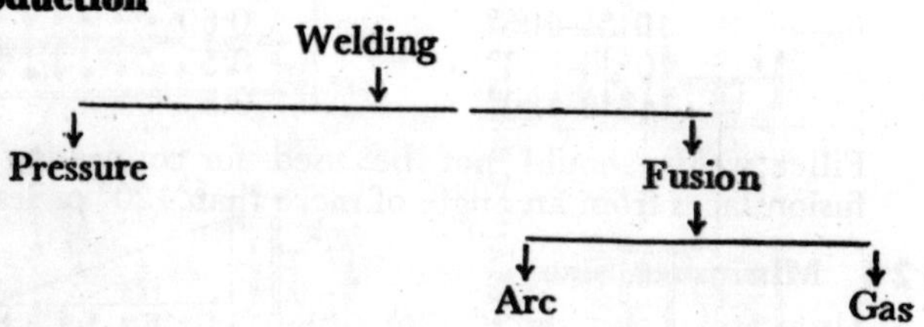

10·2 Advantages

It is modern method of welding
Economy—Elimination of connecting angles, plates, rivet heads, guesset, splice plates—labour
Stiffness—Connections are rigid
Efficiency—100% can be achieved
Others—Noiseless operation, clean and neat joints. Dead load reduction

10·2·1 Disadvantages

(*i*) Over rigidity (*ii*) Warping
(*iii*) Fatigue (*iv*) Brittle
(*v*) X-ray testing and inspection—skilled labour

10·2·2 Types

(*i*) Butt welds—Single V, U, J
(*ii*) Fillet welds—Double V, U, J

10·2·3 Definitions

Fillet welds—A weld approximately traingular in cross-section joining two surfaces at right angles to each other in a lap, tee or corner joint.
Leg—Fusion face in a fillet weld
Leg length—The distance from the root to the toe of a fillet weld, measured along the fusion face.

10·2·4 Size

The size of the weld shall be specified by its effective throat thickness.

10·2·5 Effective Throat Thickness

In the case of a complete penetration on butt weld, the effective throat thickness shall be taken as the thickness of the thinner part joined

In the case of a fillet weld, the effective throat thickness shall be taken as :

K times fillet size where *K* is a constant which depends on the angle between the fusion faces.

Angle	*K*
90°—60°	0·7
91°—100°	0·65
101°—106°	0·60
107°—113°	0·55
114°—120°	0·5

Fillet welds should not be used for connecting parts whose fusion faces from an angle of more than 120° or less than 60°.

10·2·6 Minimum Size

Upto and including 9·5 mm	3 mm size
Over 9·5 mm including 19 mm	5 mm
Over 19 mm—32 mm	6 mm
Over 32 mm—50 mm	8 mm
Over 50 mm	Special precautions are necessary

10·2·7 Permissible Stress Butt Welds

Direct tension		1420 kg/cm²
Direct comp		1420 kg/cm²
In bending		
Tension	1575 kg/cm²	
Comp	1575 kg/cm²	
In shear	1025 kg/cm²	

10·2·8 Site Welds

80% of above stress

10·2·9 Increase in Stresses

When wind loads are considered—an increase of 25% is allowed

10·2·10 Minimum Length

The effective length of a fillet weld designed to transmit loading shall not be less than four times the size of the weld or 3·8 cm.

10·2·11 Spacing

30 cm or 16 × thickness of the thinner part in compression or 24 × thickness of the thinner part in tension

End returns	= 2 × size of the weld
Size of the weld	= S
Effective size	= $0·7\ S$
Effective length of weld	= l
Allowable stress in shear	= 1025 kg/cm²
Strength of weld	= $0·75 \times l \times 1025$

10·3 Welding Symbols

Bead ...

Fillet ...

Square ...

V ...

Bevel ...

U ...

J ...

Plug or slot ...

Field weld ...

Weld around...

Flush ...

11

Simple Welded Connection

11·1 Problem

A welded plate girder is made up of 40×2.5 cm flange plates and 160×1 cm web plate. The beam resists a maximum shear force of 80 tonnes. Design the joint between the flange and web plates using intermittent shop fillet welds.

11·2 Allowable Stress

Shear stress=1025 kg/cm^2 for shop fillet welds.

11·2·1 Minimum Size of Weld

Minimum size of weld for plate thickness of **10 mm** $\nless$ 5 mm

11·2·2 Maximum Size of Weld

1·5 mm less than the normal thickness of plate

$=10-1\cdot5=8\cdot5$ mm

Use 6 mm fillet welds.

11·2·2 Effective Length of Weld

The effective length $\nless$ 4 times the weld size or 38 mm whichever is greater.

$\nless 6\times4=24$ mm

$\therefore\quad \nless 38$ mm

11·2·4 Spacing

Spacing between the effective lengths of intermittent fillet welds $\ngtr$ $16t$ for compression

$\ngtr$ $24t$ tension

$\ngtr$ 30 cm.

where t=thickness of the thinnest plate in joint

11·2·5 Shear Force

Shear Force=F=80 tonnes.

Shear stress at the junction of I section

$$q=\frac{F_x\,A\bar{y}}{IZ}$$

$$I_x=\frac{1}{12}\times1\times160^3+2\times40\times2\cdot5\times81\cdot25^2$$

$$+\frac{1}{12}\times40\times2\cdot5^3\times2=49000+1300000+104=1349104\text{ cm}^4$$

$A\bar{y}=40\times2\cdot5\times81\cdot25=8125$ cm^3

Z=Effective thickness of weld section=$2\times$thickness

$=2\times{\cdot}6\times0{\cdot}707={\cdot}845$ cm

$$q=\frac{80000\times8125}{1349104\times{\cdot}845}=56{\cdot}5\text{ kg/cm}^2$$

11·2·6 Strength of Weld

Strength of weld per cm length of weld $=0.707\times{\cdot}6\times1025$
$=432$ kg

Welded length required per metre length of beam

$$\frac{56{\cdot}5}{432}\times100=13{\cdot}1\text{ cm}$$

Use 5 mm—5 cm long welds with minimum spacing of
$16t=16$ cm in the compression
and $24t=24$ cm in the tension flanges

11·2·7 Arrangement of Welds

5 mm size 5 cm in length

Spacing in compression flange=15 cm

Spacing in tension flange on either side of the web=20 cm

12

Welded Crane Bracket

12·1 Problem

A crane bracket carries a load of 10 tonnes at a distance of 15 cm from the edge of column ISWB 300

Plate thickness of the bracket=15 mm

Design the joint.

12·2 Data

ISWB 300—b=20 cm $\qquad t_f$=9·9 mm

12·2·1 Trial Arrangement of Weld

Min : size of weld=5 mm for plate above 9 mm

Maximum size of weld=9·9−1·5=8·4 mm

∴ Use 6 mm fillet weld.

Approximate Length required to resist direct load on one flange of column alone$=\frac{10\times1000}{2\times0·6\times·707\times1025}=11·7$ cm

Try 60 cm long fillet weld in order to resist torsional stress due to bending

12·2·2 Centre of Gravity of Weld Section

Taking moments about section 1—1

$$60\times\bar{x}=2\times15\times\frac{15}{2}$$

$$\bar{x}=\frac{2\times15\times15}{60\times2}=15/4=3·75 \text{ cm}$$

Eccentricity of the load=15+15−3·75=30−3·75=26·25 cm

Moment due to eccentric load$=\frac{10}{2}\times26·25=131·25$ cm tonnes

=131250 cm kg

12·2·3 Polar Moment of Inertia

$$I_{xx}=\frac{1}{12}\times0·707\times·6\times30^3+2\times0·707\times·6\times15\times15^2$$

$$=950+2860=3810 \text{ cm}^4$$

$$I_{yy}=2\left(\frac{1}{12}\times0·6\times·707\times15^3+0·707\times·6\times25\times3·75^2\right)$$

$$+30\times·6\times·707\times3·75^2$$

$$=2(120+149)+178=538+178=716 \text{ cm}^4$$

$$J=I_{xx}+I_{yy}=3810+716=3526 \text{ cm}^4$$

Maximnm stress in torsion will occur at points —2

12·2·4. Stresses

Direct stresses

$$f_a=\frac{10000}{2\times60\times·6\times·707}=186 \text{ kg/cm}^2$$

$$r=\sqrt{11·25^2+15^2}=\sqrt{126+225}=\sqrt{351}=18·7 \text{ cm}$$

Torsional stress$=\frac{Tr}{J}$, $f_m=\frac{131250\times18·7}{3526}=695$ kg/cm²

$$\text{Cos }\theta=\frac{11·25}{18·7}=0·605$$

Resultant stress $= \sqrt{f^2_a + f^2_m + 2 f_a f_m \cos\theta} = \sqrt{519700 + 78500}$

$= \sqrt{598200} = 785 \text{ kg/cm}^2 < 1025 \text{ kg/cm}^2$

Hence the trial arrangement is safe.

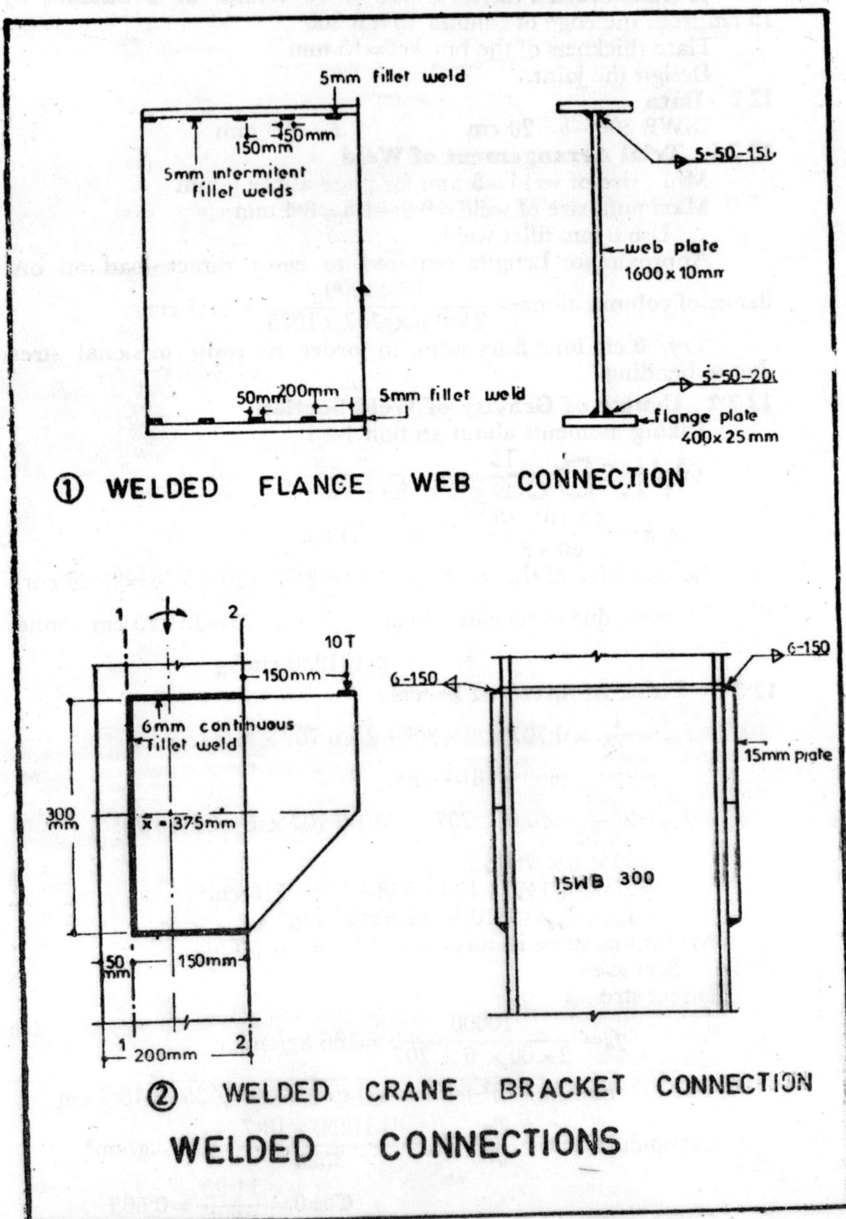

WELDED CONNECTIONS

VI. Design of Beams

1

Beams–General

1·1 Definition

A beam is a structural element or member subjected to transverse loads and reactions.

A large beam supporting smaller beam is a girder.

1·2 Design

Designing a beam is meant the determination of minimum section for the beam to be able to resist the applied loads safely. A section heavier than the minimum section will naturally be able to carry the loads quite safely but will not be economical.

A design should therefore be in consistence with safety and economy.

1·3 Data

(1) The maximum moments and shear forces and the points at which these occur.

(2) The position of zero moment termed the points of contra-flexure or the point of inflextion.

(3) Allowable stresses in the material of the beam in tension, compression, shear and bending and its modulus of elasticity.

(4) Properties of sections such as section modulus, area, and weight.

1·4 Bending Stresses

For Rolled I beams with or without cover plates and double channel beams with cover plates when compression flange is laterally supported.

Rolled I-beams and channels (tension or compression)

$$=1650 \text{ kg/cm}^2$$

For other Rolled sections and plate girders

$$f_t=1575 \text{ kg/cm}^2$$

1·5 Allowable Stresses

Axial Stresses

In tension $f_t=1500$ kg/cm²

In compression f_c as given in the Table.

1/r	Allowable Axial Stress in kg/cm² f_a	1/r	Allowable Axial Stresses in kg/cm² f_c
0	1250	170	377
10	1246	180	336
20	1239	190	300
30	1224	200	270
40	1203	210	243
50	1172	220	219
60	1130	230	199
70	1075	240	181
80	1007	250	166
90	928	300	109
100	840	350	76
110	753		
120	671		
130	597		
140	531		
150	474		
160	423		

1·6 Shear Stresses

For parts other than Rolled steel sections maximum shear stress $f_s = 1100$ kg/cm²

For webs unstiffened d/t not greater than 85

where d = clear distance between flanges and
t = thickness of web

Average shear stress = 945 kg/cm²

1·7 Live Loads on Buildings

LIVE LOADS ON FLOORS

Loading Class	Type of Floors	Live Loads kg/cm²
200	Floors in dwelling houses, tenements, hospital wards, bed rooms and private sitting rooms in hostels, and dormitories	200
250	Office floors other than entrance halls, floors of light workrooms	250—400
300	Floors of banking halls, office entrance halls and reading rooms	300
400	Shop floors used for the display and sale of merchandise, floors of work rooms generally floors of class rooms and restaurants	400

500	Floors of warehouses, workshops, factories and other buildings or parts of building of light weight loads	500
750	Floors of warehouses, workshops, factories and other buildings or parts of buildings of similar category for medium weight loads	750
1000	Floors of warehouses, workshops, factories other buildings for heavy weight loads, floors of book stores and libraries	1000
Gara e (Light)	Floors used for garage for vehicles not exceeding 2·5 metric tonnes gross-weight :	
	Slabs	400
	Beams	250
Garage (Heavy)	Floors used for garages for vehicles not exceeding 4 metric tonnes gross-weight	750
Stairs	Stairs, landings and corridors for class 200 loading but not liable to overcrowding.	300
	Stairs. landings and corridors for class 200 loading but liable to overcrowding and for all other classes.	500
Balcony	Blaconies not liable to overcrowding :	
	For class 200 loading	300
	For all other classes	500
	Balconies liable to overcrowding	500

These are applicable for slabs more than 2·5 m span and beams supporting 6 m² of floor. For smaller spans of slab the minimum load on the full span should be 8 times the above value. For beams supporting areas of slab less than 6 sq. metres should be designed for 64 times the above value.

For light service buildings, the design of columns, walls and foundations need not allow for full live loads of the floors above and reduction may be allowed as follows :

No. of floors supported by the member as in Question	% of reduction in imposed load of all floors above member
1	0
2	10
3	20
4	30
5	40
and above	40

6th	floor	40%
5th	floor	40%
4th	floor	30%
3rd	floor	20%
2nd	floor	10%
1 st	floor	0%

DEAD LOADS

Item	Dead Load
Reinforced Concrete	2400 kg/m³
Plain Concrete	2300 kg/m³
Sand	1600 kg/m³
Granite Stone	2700 kg/m³
Timber	700 kg/m³
Brick Work	1900 kg/m³
Tiles	70 kg/m³

1·8 Deflection

The maximum deflection for beams should not be more than 1/325 of span.

For cantilevers 2/325 of span.

1·9 Choice of the Section

Though two beams of different depths have the same section modulus, the suitability or the choice is governed by the weight of the section, deflection of beam. Deeper sections give lesser deflection.

It is therefore suggested for the required section modulus, the deepest section which is also lightest in weight should be chosen.

DESIGN PROCEDURE

Various steps in the design of a beam are as follows :

1. Calculate the dead and live loads that the beam has to support. Draw a diagram showing the beam with all loads and reactions.
2. Calculate the Max. B.M. and S.F.
3. Calculate section modulus $M/f=Z$.

4. Select a section (deeper and lighter) slightly more than what is required by calculations.
5. Test the section for shear and deflection.
6. Depth (D) of section should not be less than $L/24$ if so, for $D > L/24$ the section satisfies the requirements the deflection of 1/325 of span.

1·10 Modulus of section

$$\frac{M}{I} = \frac{f}{y} \text{ from theory of simple bending}$$

or $$M = f\,\frac{I}{y} = fZ \text{ where } Z = \text{section modulus}$$

or $$Z = \frac{M}{f} = \frac{\text{Moment}}{\text{Allowable stress}}.$$

1·11 Radius of Gyration

$$I = Ar^2 \text{ or } r = \sqrt{I/A}$$

$$r_{xx} = \sqrt{\frac{I_{xx}}{A}},\quad r_{yy} = \sqrt{\frac{I_{yy}}{A}}$$

10·12 Shearing Stresses

Shearing Stress $q = \dfrac{FA\bar{y}}{Ib}$

where
F = shear force on the section
I = Moment of Inertia
b = Width of section at the fibre at which the shear is required.
q = Shear stress
$A\bar{y}$ = Moment of area of the section situated beyond the fibre away from N.A.

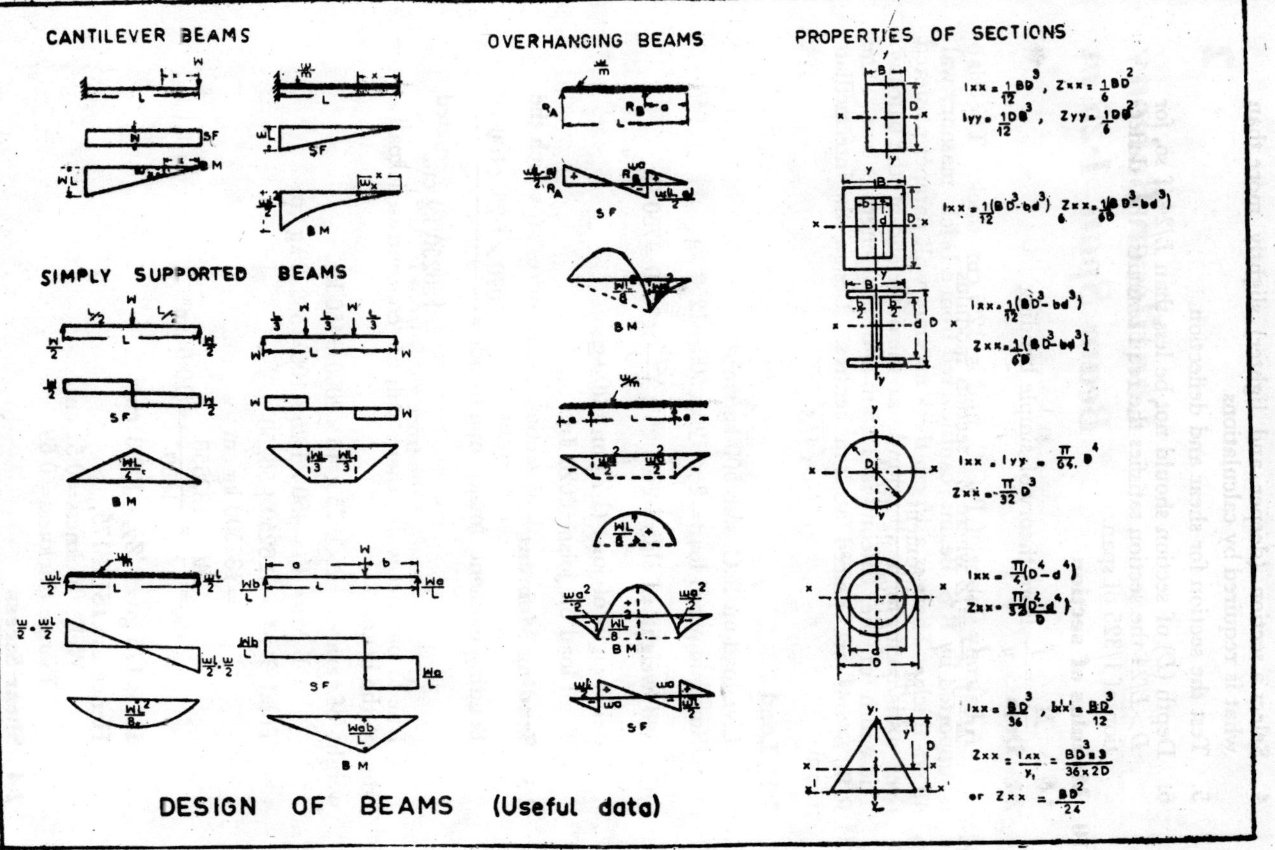
CANTILEVER BEAMS
OVERHANGING BEAMS
PROPERTIES OF SECTIONS
SIMPLY SUPPORTED BEAMS
SF
BM
DESIGN OF BEAMS (Useful data)

2

Cantilever Balcony Beam Span–1·25 m

2·1 Data

A balcony slab is of R.C.C. and is 12·5 cm thick. The slab is supported by R.S. Beams cantilevered from a brick masonry wal 45 cm thick. There is a 15 cm thick masonry wall parapet plastered over and 75 cm high all round at the outer end of the balcony. The width of the balcony is 1·25 m and the length is 8 m. The R.S. Beams are placed at 2 m centres. Design the intermediate beam.

2·2 Load

Live load on R.C. slab 500 kg/cm².

Super imposed load$=2\times1{\cdot}25\times500=1250$ kg.

Self weight of slab$=2\times1{\cdot}25\times\frac{12{\cdot}5}{100}\times2400=750$ kg.

Self weight of joist (1·25 m)$=20$ kg

Total load on joist$=2020$ kg.

2·3 Bending Moment

Bending moment (max) due to udl$=\frac{2020\times1{\cdot}25\times100}{2}$

$=126250$ kg cm.

Due to parapet wall, there will be concentrated load at the edge of the beam

weight of parapet$=2\times0{\cdot}75\times{\cdot}15\times2000=450$ kg

B.M.$=Wl=450\times1{\cdot}25\times100=56250$ kg cm

Total B.M.$=126250+56250$

$=182500$ kg cm

$$Z_{xx}=\frac{M}{f}=\frac{182550}{1650}=110{\cdot}6 \text{ cm}^3$$

ISLB 175 gives $Z_{xx}=125{\cdot}3$ cm³

Hence use ISLB 175,

Web thickness$=0{\cdot}51$ cm

Flange thickness$=0{\cdot}69$

2·4 Shear Stress

Max. shear$=2020+450=2470$ kg

Area of web $=0{\cdot}51\ (17{\cdot}5-2\times0{\cdot}69)=0{\cdot}51\ (17{\cdot}5-1{\cdot}38)$
$=0{\cdot}51\times16{\cdot}12=8{\cdot}22\ \text{cm}^2$

Average shear $=\dfrac{S}{A}=\dfrac{2470}{8{\cdot}22}=300\ \text{kg/cm}^2<945\ \text{kg/cm}^2$

2·5 Deflection

Permissible deflection $=\dfrac{1}{355}\times1{\cdot}25\times1000=3{\cdot}84$ mm

$D>L/24=\dfrac{1{\cdot}25\times100}{24}=5{\cdot}22$ cm

Provided depth = 17·5 cm

Hence $D>\dfrac{L}{24}$ Hence safe.

$I_{xx}=1096{\cdot}2\ \text{cm}^4$, $E=2\times10^6\ \text{kg/cm}^2$

$$\text{Deflection}=\frac{w_1\ l^3}{8\ EI}+\frac{w_2\ l^3}{3\ EI}$$

$$=\frac{2020\times125^3}{8\times2\times10^6\times1096{\cdot}2}+\frac{540\times125^3}{3\times2\times10^6\times1096{\cdot}2}=0{\cdot}108+0{\cdot}064$$

$=0{\cdot}172$ cm or $1{\cdot}72$ mm $<3{\cdot}34$ mm

Hence the section ISLB 175 is sufficient.

3

Lintels

3·1 Introduction

Lintels are small beams provided over the door or window openings to support the masonry walls.

These small beams will be usually of R.C.C. and structural steel sections will be rarely used. Since there is some action in the masonry it is assumed that the lintel carries a load of triangular shape shown in figure. The angle at the base is assumed 45° for good masonry and 60° for poor masonry.

Arch action is possible only when the wall above the lintel is not less than the height of the triangle, and minimum height of the wall should be $1\frac{1}{4}$ h.

The lintel has to be designed for a triangular load if the wall above lintel is greater than $1\frac{1}{4}$ h where "h" is the height of the triangle, if the height is less than $1\frac{1}{4}$ h, then the lintel has to be designed for complete load of full length of masonry.

4

Lintel Beam Span–4 m

4·1 Data

An opening of 3·7 m clear span is to be made in a well-bonded brick wall 40 cm thick. The lintel beam for the opening, in addition to carrying the brickwork loading, has to support joints which transmit a total uniform load of 20 tonnes.

4·2 Loads

Assuming 30 cm bearing, effective span=3·70+·15+·15=4m

The wt of brickwall triangular shape (60°)

$$w_1=\frac{1}{2}\times 4\times\frac{\sqrt{3}}{2}\times 4\times\frac{40}{100}\times 1920=3072\sqrt{3}\text{ kg}$$

=5320 kg or 5·32 tonnes

Total load W=5·32+20=25·32 tonnes

4·2 1 BM

$$\text{Max. B.M}=\frac{Wl}{6}=\frac{25{\cdot}32\times 4\times 1000\times 100}{6}=1688000\text{ cm kg}$$

B.M. due to self wt $\left(\text{assuming } \frac{L}{10}=40 \text{ cm for heavy loads}\right)$ at 57 kg/m

$$=\frac{57\times 4\times 4\times 100}{8}=11400\text{ cm kg}$$

Total B.M.=1688000+11400=1699400 cm kg

$$Z_{xx}=\frac{M}{f}=\frac{1699400}{1650}=1030\text{ cm}^3$$

4·2·2 Selection of Section

ISLB 450 weighing 65·3 kg/m gives Z_{xx}=1223·8 cm³

$$\text{B.M. due to self wt}=\frac{65{\cdot}3\times 4\times 4\times 100}{8}=13060\text{ cm kg}$$

Revised B.M.=B.M. due to triangular load+B.M. due to sel wt of beam =1688000+13060=1701060 cm kg

$$\text{Revised } Z_{xx}=\frac{M}{f}=\frac{1701060}{1650}=1035\text{ cm}^3$$

ISLB 450 gives Z_{xx}=1223·8 cm³

Use ISLB 450

4·2·3 Shear Stress

$$\text{Max. shear}=\frac{25{\cdot}32}{2}+\frac{65{\cdot}3\times 4}{2\times 1000}=12{\cdot}66+{\cdot}013$$

=12·673 tonnes

Area of web $=t\,(d-2T)$=0·86(45−2×1·34)

=0·86 (45−2·68)=0·86×42·32=36·4 cm²

$$\text{Average Shear}=\frac{12673}{36{\cdot}4}=346\text{ kg/cm}^2 \not< 945\text{ kg/cm}^2$$

(Hence safe)

5

Girder for an Office Floor Span–12 m

5·1 Data

A floor of an industrial building to be supported on the beams and girder as shown in the sketch, and it is expected to carry a 12·5 cm. thick slab of dead load 300 kg/m² and a live load of 500 kg/m².

Design a cross joist and main girder.

5·2 Load Beam B_1

Span for Beam B_1 =4 m
Dead load =300×3=900 kg/m
Live load =500×3=1500 kg/m
Self wt (25 cm depth)

$$\frac{1}{15} \text{ to } \frac{1}{20} \text{ (span)}=28 \text{ kg/m}$$

w=Total load=2428 kg/m

3·2·1 Bending Moment

$$\text{B.M.} = \frac{wl^2}{8} = \frac{2428\times4\times4\times100}{8} = 485600 \text{ cm/kg}$$

$$Z_{xx} = \frac{M}{f} = \frac{485600}{1650} = 294 \text{ cm}^3$$

ISLB 250 gives Z_{xx}=297·4>294 cm³
Use ISLB 250

5·2·2 Shear Stress

Area of web=0·61 (25−2×0·82)=0·61 (25−1·64)
=0·61×23·36=14·2 cm²

$$\text{Max Shear} = \frac{wl}{2} = \frac{2428\times4}{2} = 4856 \text{ kg}$$

$$\text{Average Shear Stress} = \frac{4856}{14\cdot2} = 342 \text{ kg/cm}^2$$

342 kg/cm² < 945 kg/cm² (safe)

5·3 Beam Loads B_2

Span=12 m

Dead load	=3×4×300=3600 kg	
Live load	=3×4×500=6000 kg	
Self weight	=4×28	=112 kg
	Total load	=9712 kg

Reaction from secondary beams B_1 on the main girder

B_2=9712 kg

5·3·1 Bending Moment

Bending moment due to the concentrated reactions from secondary beams$=\frac{3}{2}\times9712\times6\times100-9712\times3\times100=8740800$
$-2913600=5827200$ cm kg

Bending Moment due to self wt $\left(\frac{L}{20}=\frac{1200}{20}=60\text{ cm, at }145{\cdot}1\text{ kg/m}\right.$

$$=\frac{145{\cdot}1\times12\times12\times100}{8}=261180\text{ cm kg}$$

Total B.M.$=5827200+261180=6088380$ cm kg

$$\text{Max Shear}=\frac{3}{2}\times9712+\frac{12\times145{\cdot}1}{2}=14700+871$$

$$=15571\text{ kg or }15{\cdot}571\text{ tonnes.}$$

$$Z_{xx}=\frac{M}{f}=\frac{6088380}{1650}=3685\text{ cm}^3$$

ISWB 600 gives $Z=3854{\cdot}2\text{ cm}^3>3685\text{ cm}^3$

Use ISWB 600

$t=1{\cdot}18$

$T=2{\cdot}36$

5·3·2 Shear Stress

Area of web $=t\,(d-2T)$

$=1{\cdot}18(60-2\times2{\cdot}36)$

$=1{\cdot}18\,(60-4{\cdot}72)$

$=1{\cdot}18\times55{\cdot}28$

$=65\text{ cm}^2$

$$\text{Average shear}=\frac{15571}{65}=238\text{ kg/cm}^2$$

$$<945\text{ kg/cm}^2$$

$$\text{Deflection}=\frac{19}{384}\times\frac{wl^3}{EI}$$

or $$d>\frac{L}{24}=\frac{12\times100}{24}=50\text{ cm}$$

Provided depth=60 cm.

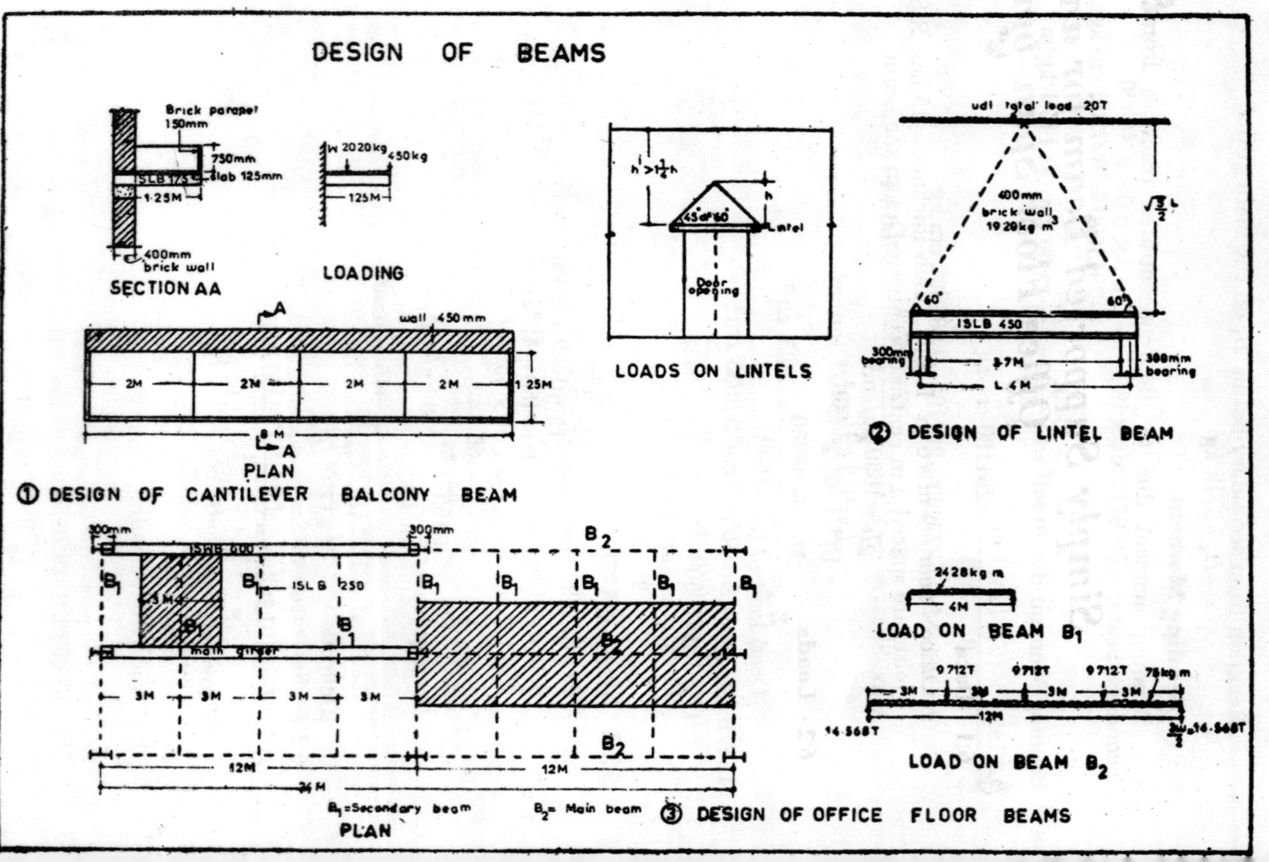
DESIGN OF BEAMS
Brick parapet 150mm
750mm
slab 125mm
1·25M
400mm brick wall
SECTION AA
2020kg
450kg
125M
LOADING
wall 450mm
2M
8 M
PLAN
① DESIGN OF CANTILEVER BALCONY BEAM
Door opening
Lintel
LOADS ON LINTELS
udl total load 20T
400mm brick wall 1920kg m³
ISLB 450
300mm bearing
3·7M
L 4M
② DESIGN OF LINTEL BEAM
300mm
ISLB 250
main girder
3M
12M
24M
B1=Secondary beam
B2= Main beam
PLAN
2428kg m
4M
LOAD ON BEAM B1
9·712T
75kg m
14·568T
LOAD ON BEAM B2
③ DESIGN OF OFFICE FLOOR BEAMS

6

Simply Supported Beam for an Office Floor Span–6m

6·1 Data

Office floor 9 m×6 m R.C.C. Slab-thickness=15 cm R.S. Joists are spaced 3 m centre to centre. Design the beam

$f_b=1650$ kg/cm^2

$f_s=945$ kg/cm^2

6·2 Loads

Dead load of R.C. Slab$=\frac{15}{100}\times 1\times 2400=360$ kg/m^2

Live Load on R.C. Slab =250 kg/m^2

Floor finishing =40 kg/m^2

Total=650 kg/m^2

Load on each girder$=3\times 6\times 650$ =11700 kg

Depth 1/20 span$=\frac{1}{20}\times 6\times 100$ =30 cm

Self wt of beam (ISLB−300)$=37{\cdot}7\times 6$ =226 kg

Total Load on Beam=11700+226 =11926 kg

$$BM\text{ Max.}=\frac{Wl}{8}=\frac{11926\times 6\times 100}{8}$$

$$=11926\times 75=895000 \text{ cm kg}$$

$$Z_{xx}=\frac{M}{f}=\frac{895000}{1650}=543 \text{ cm}^3$$

6·2·1 Selection of Section

Try ISLB 325 section, weight 43·1 kg/m

$Z_{xx}=607{\cdot}7$ cm^3 > 543 cm^3

Additional self weight of beam$=5{\cdot}4\times 6=32{\cdot}4$ kg

Additional Bending Moment $=\frac{32{\cdot}4\times 6\times 100}{8}$

$=32{\cdot}4\times 75=2430$ cm kg

Revised BM=895000+2430=897430 cm kg

$$Z_{xx}=\frac{M}{f}=\frac{897430}{1650}=545 \text{ cm}^3$$

ISLB 325 gives 607·7 cm^3 > 545 cm^3

Use ISLB 325.

6·2·2 Shear Stress

$$\text{Max. shear} = \frac{11958{\cdot}4}{2} = 5979{\cdot}2 \text{ kg}$$

$$\text{Area of the web} \quad t=(d-2T)$$
$$=0{\cdot}70(32{\cdot}5-2\times 0{\cdot}98)$$
$$=0{\cdot}70(32{\cdot}5-1{\cdot}96)$$
$$=0{\cdot}7\times 30{\cdot}54$$
$$=21{\cdot}378 \text{ cm}^2$$

$$\text{Average shear stress} = \frac{5979{\cdot}2}{21{\cdot}378} = 284 \text{ kg/cm}^2$$
$$\not< 945 \text{ kg/cm}^2 \text{ (safe)}$$

6·2·3 Deflection

$$d > L/24$$

$$d = \frac{6\times 100}{24} = 25 \text{ cm}$$

Provided 32·5 cm depth. Hence safe.

7

Beams for an Office Floor

7·1 Data

An office building floor consists of rooms 7·5 m wide on one side of the corridor 3 m wide and 4·5 m wide rooms on the other side as shown in plan. The columns are spaced at 3 m centre to centre. Slab thickness for room floor is 12 cm and for corridor it is 10 cm thick. Design the Beams B_1, B_2 and B_3

7·2 Beam, B_1

Span $l=4\cdot5$ m

Beam supports 3 m wide RCC floor slab.

Loads

Loads per metre length of beam

Dead load of RCC slab (12 cm thick)

$$=\frac{12}{100}\times1\times3\times2400=864 \text{ kg}$$

Live load at 400 kg/m for office floor $=3\times1\times400=1200$ kg/m

Floor finish (40 mm thick) $=\frac{4}{100}\cdot1\times3\times2400=288$ kg/m

Self weight of beam assuming $\left(\frac{L}{20}=\frac{450}{20}\right)$

Assume ISLB 250 = 40 kg/m

Total load = 2392 kg/m

Say $w=2400$ kg/m

$$\text{Maximum Bending moment}=\frac{wl^2}{8}=\frac{2400\times4\cdot5^2\times100}{8}$$

$$=601000 \text{ cm kg}$$

$$Z_{xx}=\text{Modulus of section required}=\frac{601000}{1500}=405/\text{cm}^3$$

Use ISMB—250 @ 37·3 kg/m

Since the depth provided is greater than $\frac{L}{20}$ the deflection will be within the limits.

7.2·1 Shear Stress

$$\text{Shear Force}=\frac{2400\times4\cdot5}{2}=5400 \text{ kg}$$

Web area $= t_w \times d_w = 0{\cdot}69 \times (25 - 2 \times t_f)$
$= 0{\cdot}69(25 - 2 \times 1{\cdot}25) \cdot 69 \times 22{\cdot}5 = 15{\cdot}5$ cm²

Shear stress $= \dfrac{5400}{15{\cdot}5} = 346$ kg/cm² < 945 kg/cm²

7·2·2 Beam B₂

Span $l = 7{\cdot}5$ m

Beam support 3 m wide RCC slab

Loads

Loads per metre length of beam
Dead load of slab = 864 kg
Live load = 1200 kg
Floor finish = 288 kg

Self weight of beam $\left(\dfrac{L}{20} = \dfrac{750}{20}\right)$

Assuming ISLB 400 = 70 kg

Total load/m $w = 2425$ kg

Maximum bending moment $= \dfrac{wl^2}{8} = \dfrac{2425 \times 7{\cdot}5^2 \times 100}{8}$
$= 1710000$ cm kg

Z_{xx} = Modulus of section reqd. $= \dfrac{1710000}{1500} = 1140$ cm³

Use ISLB 450 at 65·3 kg/m ($Z_{xx} = 1223{\cdot}8$ cm³)

7·2·3 Shear Stress

Shear Force $= \dfrac{2425 \times 7{\cdot}5}{2} = 9100$ kg

Web area $= t_w \times d_w = 0{\cdot}86 \times (45 - 2 \times t_f) = 0{\cdot}86 \times (45 - 2 \times 1{\cdot}34)$
$= 0{\cdot}86 \times 42{\cdot}32 = 36{\cdot}6$ cm²

Shear stress $= \dfrac{9100}{36{\cdot}6} = 248$ kg/cm² < 945 kg/cm²

7·3 Beam B₃

Span $l = 3$ m
Beam supports, 3 m wide RCC slab

7·3·1 Loads

Dead load due to slab $= \dfrac{10}{100} \times 3 \times 1 \times 2400 = 720$ kg/m

Live load at 300 kg/m $= 300 \times 1 \times 3 = 900$ kg/m

(office entrance halls)

Floor finish (40 mm thick) $= \dfrac{4}{100} \times 3 \times 1 \times 2400 = 288$ kg/m

Self weight of beam $\left(\dfrac{L}{20}\right)$

Assume ISLB 100 = 12 kg/m
Total load $w = 1920$ kg/m

Bending moment$=\frac{wl^2}{8}=\frac{1920\times3\times3}{8}=21600$ cm kg.

Z_{xx} required $=\frac{21600}{1500}=14{\cdot}5$ cm^3

Try ISLB 75 at 6·1 kg/m

$Z_{xx}=19{\cdot}4$ cm^3

7·3·2 Shear Stress

Shear force $=\frac{1920\times3}{2}=2880$ kg

Web area $=t_w\times(d-2t_f)=0{\cdot}37\times(7{\cdot}5-2\times{\cdot}5)=0{\cdot}37\times6{\cdot}5$
$=2{\cdot}4$ cm^2

Shear stress $=\frac{2880}{2{\cdot}4}=1200$ kg/cm$^2>945$ kg/cm^2

Unsafe in shear.

Try ISLB 100,

Web area $=0{\cdot}4\times(10-2\times0{\cdot}64)=0{\cdot}4\times8{\cdot}72=3{\cdot}5$ cm^2

Shear stress $=\frac{2880}{3{\cdot}5}=820$ kg/cm$^2<945$ kg/cm^2

Use ISLB 100 at 8 kg/m.

$Z_{xx}=33{\cdot}6$ cm^3.

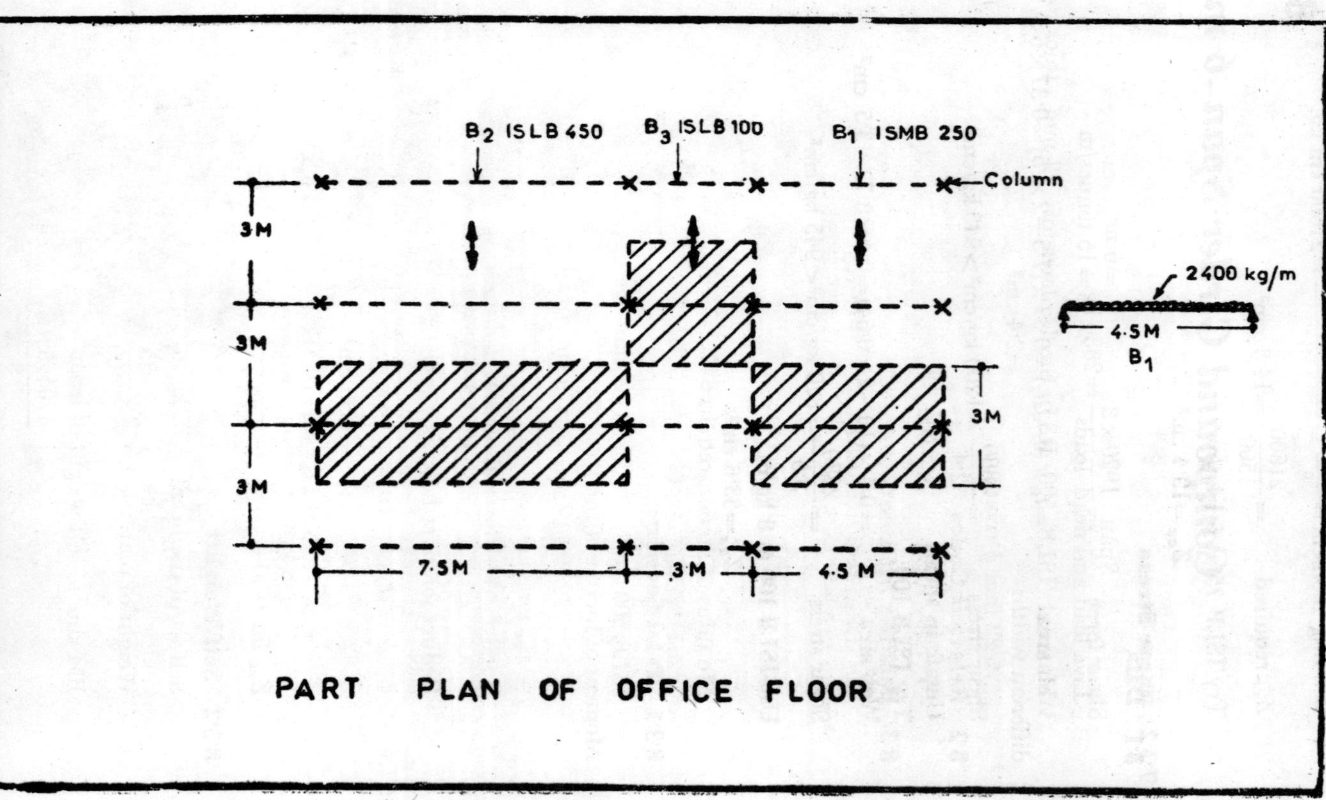

PART PLAN OF OFFICE FLOOR

8

Compound Girder Span–6 m

8·1 Data

Span $=6$ m
Live load and dead loads $=15$ tonnes/m

Material ISLB 500 RSJ, Steel plates of 12 mm thick of different widths

22 mm ϕ rivets.

8·2 Relevant Codes

IS—875, IS—800

8·3 Bending Moment

Dead and live loads $=15$ tonnes/m
Span $=6$ m

$$\text{Bending moment } M=\frac{15\times6\times6}{8}=67.5 \text{ m tonnes}$$

$$\text{Modulus of section required } Z_{xx}=\frac{67{\cdot}5\times100}{1{\cdot}65}=4080 \text{ cm}^3$$

8·3·1 Trial Section

ISLB 500 with 2–250×12 mm cover plates on each flange.
Moment of inertia of the section

$$I_{xx} \text{ for beam}=38579 \text{ cm}^4$$

$$I_{xx} \text{ for plates } Ah^2=2\times25\times2{\cdot}4\times26\ 2^2=86100 \text{ cm}^4$$

Total I_{xx} for the trial section

$$=86100+38579=124679 \text{ cm}^4$$

Deduct for rivet holes in the flanges.

$$I_{xx} \text{ for holes}=4\times2{\cdot}35\times3{\cdot}81\left(27{\cdot}4-\frac{3{\cdot}81}{2}\right)^2$$

$$=8850 \text{ cm}^4$$

$$\text{Net } I_{xx}=124679-8850=115829 \text{ cm}^4$$

$$Z_{xx} \text{ for trial section } =\frac{115829}{27{\cdot}4}=4210 \text{ cm}^3$$

8·3·2 Self Weightt

Self wt of the beam $=75$ kg/m

$$\text{Weight of plates } = 2\times\frac{25}{100}\times\frac{2{\cdot}4}{100}\times7850\times1=94{\cdot}5 \text{ kg/m}$$

BM due to self weight of beam

$$=\frac{170\times6\times6}{8}=765 \text{ m kg}$$

Addition Z_{xx} reqd. $=\dfrac{76500}{1650}=46{\cdot}1\ \text{cm}^3$

Total Z_{xx} required for the section due to all loads

$=4080+46{\cdot}1=4126{\cdot}1\ \text{cm}^3$

Z_{xx} for the trial section

$=4210\ \text{cm}^3$

Hence use ISLB 500 with two cover plates of 250×12 mm on each flange.

8·3·4 Shear Stress

$$\text{Maximum SF}=\frac{15\times6\times1000}{2}+\frac{170\times6}{2}$$

$$S=45510\ \text{kg}$$

$$\text{Shear stress}=\frac{S}{t_w\times d}=\frac{45510}{0{\cdot}92\times(50+2{\cdot}4\times2)}$$

$$S=\frac{45510}{54{\cdot}8\times{\cdot}92}=910\ \text{kg/cm}^2<945\ \text{kg/cm}^2.$$

8·3·5 Connection

Vertical shear/cm depth

$$=\frac{\text{Shear force}}{\text{Depth of web}}$$

$$=\frac{45510}{(50-2\times1{\cdot}4)\times1000}$$

$$=\frac{45510}{47{\cdot}18\times1000}=0{\cdot}965\ \text{tonnes.}$$

Horizontal shear at the flange level=965 kg.

Use 22 mm ϕ.

Strength of rivet in double shear

$$=2\times\frac{\pi\times2{\cdot}35^2}{4}\times945=8{\cdot}2\ \text{tonnes.}$$

Strength in bearing $=1{\cdot}2\times2{\cdot}125\times2{\cdot}35$

$=6$ tonnes.

Spacing of rivets $=\dfrac{6000}{965}=6{\cdot}2$ cm c/c

Maximum spacing of rivets

$=10t=10\times1{\cdot}2=12$ cm c/c

Minimum spacing $=2{\cdot}5\ d=2{\cdot}5\times2{\cdot}2=5{\cdot}5$ cm

Therefore use 22 mm ϕ at 6 cm c/c.

VII. Design of Columns And Foundations

I

Steel Stanchion with Grillage Foundations

1·1 Data

Total axial load on the section =200 tonnes
Length of column =5 m

Column is effectively held in position at both ends and restrained in direction at one end

Compressive stress in concrete $=400$ tonnes/m^2
Bearing capacity of soil $=30$ tonnes/m^2

Design :

(*i*) Steel column
(*ii*) Lacing
 (*a*) Single lacing system
 (*b*) Double lacing system
 (*c*) Battening
(*iii*) Column base
 Slab base
 Gussetted base
(*iv*) Grillage foundation
(*v*) Column cap.

1·2 Relevant Codes

IS—800

1·3 Column Trial Section

Load on the column =200 tonnes
Length of column =5 m
Assuming approximate allowable stress $=1200$ kg/cm^2

Gross area required$=\dfrac{200\times 1000}{1200}$ $=167$ cm^2

Try two ISWB 400 RSJ beams
Area provided$=85{\cdot}01\times 2=170{\cdot}02$ cm^2
For ISWB 400

$r_{xx}=16{\cdot}60$ cm, $r_{yy}=4{\cdot}04$ cm

For strength and economy for a stanchion, the two beams should be arranged such that, $r_{xx}=r_{yy}=r_{min}$.

1·3·1 Effective Length

For the end conditions given

$l_e=0{\cdot}85\ L=0{\cdot}85\times 5=4{\cdot}25$ m$=425$ cm

$r_{min}=16{\cdot}60$ cm

$$\frac{l_e}{r_{min}}=\frac{425}{16{\cdot}6}=25{\cdot}6$$

1·3·2 Permissible Stress

for $\frac{l}{r}=20,\ p_c=1239$ kg/cm²

$\frac{l}{r}=30,\ p_c=1224$ kg/cm²

By interpolation, for $\frac{l}{r}=25{\cdot}6,\ p_c=1230{\cdot}6$ kg/cm²

1·3·3 Safe Load

Safe load on the column $=p_c\times A=1230{\cdot}6\times170{\cdot}02$
$=210{,}000$ kg $=210$ tonnes.

1·3·4 Spacing of the Beams

In order to have the radius of gyration $I_{xx}=I_{yy}$, the beams should be placed at a suitable distance.

For one ISWB 400
$I_{xx}=23436{\cdot}7$ cm⁴, $I_{yy}=1388$ cm⁴

For two beams spaced at distance x apart
$I_{xx}=2\times23436{\cdot}7=46873{\cdot}4$ cm⁴

$$I_{yy}=2\left[1388+85{\cdot}01\left(\frac{x}{2}\right)^2\right]=2776+\frac{170{\cdot}02\times x^2}{4}$$

If $I_{xx}=I_{yy}$, $46873{\cdot}4=2776+\frac{170{\cdot}02\ x^2}{4}$

$$42{\cdot}5\ x^2=44097{\cdot}4$$
$$x=32{\cdot}3 \text{ cm}$$

Therefore use trial section consisting of 2—ISWB 400 RS Joists spaced at 32·5 cm c/c of webs.
This section carries a safe load $=210$ tonnes >200 tonnes.

1·4 Lacing—Single System

Try flats as lacing bars
Angle of inclination shall be between 40° and 70°

1·4·1 Minimum Inclination

$\frac{l}{r}$ for the column $=25{\cdot}6$

Maximum slenderness ratio of the lacing bar
$=0{\cdot}7\times25{\cdot}6=18$

Assuming that the rivets are used at an edge distance of 5 cm with a distance of 7 cm between the two rivets
$l_h=$ horizontal distance between two rivets
$=52{\cdot}5-5\times2=42{\cdot}5$ cm
$l_v=$ Spacing between lacing bars $=2\ l_h\cot\theta+7$
$=2\times42{\cdot}5\times\cot\theta+7=85\cot\theta+7$ cm

Slenderness ratio of lacing bar

$$=\frac{\text{Spacing of lacing bar}}{r_{min}\text{ for the component of the column}}=\frac{85\cot\theta+7}{4{\cdot}04}$$

and this should not be greater than 18

$$\therefore \quad \frac{85 \cot \theta + 7}{4{\cdot}04} = 18$$

$$85 \cot \theta = 18 \times 4{\cdot}04 - 7 = 73 - 7 = 66$$

$$\cot \theta = \frac{66}{85} = 0{\cdot}775$$

$$\theta = 52° - 12'$$

Try $\theta = 55°$, $\cot \theta = {\cdot}7002$

Spacing of the lacing bars $= 85 \cot \theta + 7 = 85 \times {\cdot}7002 + 7$
$= 59{\cdot}5 + 7 = 66{\cdot}5$ cm

1·4·2 Transverse Shear

$$S = \frac{P}{40} = \frac{200}{40} = 5 \text{ tonnes}$$

1·4·3 Force

Force in lacing bar $= \dfrac{S}{n \sin \theta} = \dfrac{5000}{2 \times \sin 55°}$

$$F_t = \frac{5000}{2 \times 0{\cdot}8191} = 3050 \text{ kg}$$

1·4·4 Width

Using 22 mm ϕ rivets minimum width required $= 6{\cdot}5$ cm.

1·4·5 Thickness

Thickness should not be less than the length between the inner end rivets of the bar /40.

Thickness $= \dfrac{42{\cdot}5 \text{ cosec } 55°}{40} = \dfrac{42{\cdot}5 \times 1{\cdot}2207}{40} = 1{\cdot}31$ cm

Try 65×14 mm flat as lacing bar

1·4·6 Effective Length of Lacing

l_e = distance between the inner end rivets of the bar
$= l_h \text{ cosec } 55°$
$= 42{\cdot}5 \times \text{cosec } 55° = 42{\cdot}5 \times 1{\cdot}2207 = 52$ cm

$$r_{min} = \frac{t}{\sqrt{12}} = 0{\cdot}405 \text{ cm}$$

$$l_e / r_{min} = \frac{52}{0{\cdot}405} = 128 < 145$$

1·4·7 Safe Load

Area of lacing bar $= 6{\cdot}5 \times 1{\cdot}4 = 9{\cdot}1$ cm^2

p_c = permissible stress for $l/r = 128$
$= 671 - (671 - 597)\frac{8}{10} = 611{\cdot}8$ kg/cm^2

Safe load $= 611{\cdot}8 \times 9{\cdot}1 = 5600$ kg $= 5{\cdot}6$ tonnes $> 3{\cdot}05$ tonnes

1·4·8 Connection

Strength of rivet in single shear $= \dfrac{\pi}{4} \times 2{\cdot}35^2 \times 1025$
$= 4450$ kg

Strength in bearing $=2{\cdot}35\times1{\cdot}4\times2360=7750$ kg
Least strength of rivet $=4450$ kg

Number of rivets needed $=\frac{3050}{4450}\simeq$ say one rivet

Use one 22 mm ϕ rivet

1·4·9 Tie-Plate

Use tie plates of 100×14 mm at the beginning and at end of lacing.

1·5 Doubling Lacing

Try flats as lacing bars
Try minimum angle of inclination $=\theta=40°$

1·5·1 Spacing

Distance between the rivet lines $=(52{\cdot}5-10)=42{\cdot}5$ cm
Spacing of lacing bars $=42{\cdot}5\cot 40°$
$=42{\cdot}5\times1{\cdot}1919=50{\cdot}4$ cm

1·5·2 Slenderness Ratio

Slenderness ratio $=\frac{50{\cdot}4}{4{\cdot}04}=12{\cdot}5$

Permissible slenderness ratio $=0{\cdot}7\times25{\cdot}6=18$

Actual slenderness ratio is less, therefore spacing of the lacing bar $=50{\cdot}4$ cm can be used

Use $l_v=50{\cdot}4$ cm with an inclination of $\theta=40°$

1·5·3 Transverse Shear

$$S=\frac{P}{40}=\frac{200}{40}=5 \text{ tonnes}$$

Force in the bar $=\frac{5}{n\sin\theta}=\frac{5}{4\times\sin 40°}$
$=1{\cdot}25\times\text{cosec } 40°=1{\cdot}25\times1{\cdot}5557=1{\cdot}94$ tonnes

1·5·4 Width

Using 22 mm ϕ rivets
Minimum width of bar $=6{\cdot}5$ cm

1·5·5 Thickness

Minimum thickness $=\frac{L}{60}$

$L=42{\cdot}5\text{ cosec } 40°=42{\cdot}5\times1{\cdot}557=66$ cm

Minimum thickness $=\frac{66}{60}=1{\cdot}1$ cm

Try a flat of 65×12 mm
$A=7{\cdot}8\text{ cm}^2$

1·5·6 Effective Length

$$r_{min}=\frac{t}{\sqrt{12}}=\frac{1{\cdot}2}{\sqrt{12}}=0{\cdot}346 \text{ cm}$$

$$l/r=\frac{46{\cdot}2}{0{\cdot}346}=133<145$$

1·5·7 Allowable Stress

$$p_c = 597-(597-531)\times\frac{3}{10} = 577{\cdot}2 \text{ kg/cm}^2$$

1·5·8 Safe Load

$$F = p_c \times A = 577{\cdot}2\times 6{\cdot}5\times 1{\cdot}2$$
$$= 7{\cdot}8\times 577{\cdot}2 = 4500 \text{ kg} > 1{\cdot}94 \text{ tonnes.}$$

1·5·9 Connection

Strength in *SS* of 22 mm ϕ rivet $=\frac{\pi}{4}(2{\cdot}35)^2\times 1025 = 4450$kg

Strength in bearing $=2.35\times 1{\cdot}2\times 2360 = 6650$ kg

Rivet value $= 4450$ kg

Number of rivets $=\frac{\text{Force}}{\text{River value}} = \frac{1940}{4450} \approx$ say 1 No.

Use 65 × 12 mm flat

1·6 Battening—Column

When battens are used, the effective length of column shall be increased by 10%

Actual effective length $= 0{\cdot}85\times 5 = 4{\cdot}25$ m

Modified effective length $= 1{\cdot}1\times 415 = 457$ cm

l/r for the column $=\frac{457}{16{\cdot}6} = 27{\cdot}5$

Permissible stress for l/r $= 27{\cdot}5$

$$= 1239-(1239-1224)\times\frac{7{\cdot}5}{10} = 1227{\cdot}75 \text{ kg/cm}^2$$

Safe load $= p_c\times A$ $= 1227{\cdot}75\times 170{\cdot}02$

$= 208$ tonnes > 200 tonnes

l_v = spacing of the battens = c/c distance of two consecutive battens

Slenderness ratio of unsupported length of column

$$= \frac{l_v}{r_{min} \text{ of the column member}} = \frac{l_v}{4{\cdot}04}$$

This should be less than either 40 or $0{\cdot}6\times l/r$ of the column whichever is the least

$$\frac{l_v}{4{\cdot}04} = {\cdot}6\times 27{\cdot}5 = 16{\cdot}5 < 40$$

$$l_v = 4{\cdot}04\times 16{\cdot}5 = 66{\cdot}5 \text{ cm}$$

Use 60 cm as the spacing of battens

1·6·1 Transverse Shear and Moment

$$S = \frac{P}{40} = \frac{200}{40} = 5 \text{ tonnes}$$

Force in the batten

$$= F = \frac{Sl_v}{\times x} = \frac{5\times 60}{2\times 32{\cdot}5} = 4{\cdot}62 \text{ tonnes.}$$

M = moment in the batten plate

$$= \frac{Sl_v}{2n} = \frac{5\times 60}{2\times 2} = 7 \text{ cm tonnes.}$$

1·6·2 Size—Depth

Minimum depth of end batten=Distance between the centroid of the member=32·5 cm

Overall depth of the end batten using 22 mm ϕ rivets for connection=32·5+2×edge distance=32·5×2×3·8=40·1 cm

Say 41 cm

Effective depth of intermediate batten=0·75×32·5=24·4 cm
Overall depth=24·4+2×3·8=32 cm

1·6·3 Thickness of the Batten

$F=t\times D\times P_q=t\times 32\times 945=4620$

$t=\cdot 152$ cm=1·52 mm

Minimum thickness $t=\frac{1}{50}\times 41=0{\cdot}82$ cm

Use 10 mm thick plate.

Moment of resistance of batten $=f\times z=1650\times t\,\frac{D^2}{6}$

$=\frac{1650\times 1\times 32^2}{6}=282{,}000$ cm kg > 75000 cm kg

Use 32 cm×1 cm flat for intermediate battens and 41×1 cm for end battens.

1·6·4 Connection

Using 22 mm ϕ rivets

Strength in single shear $=\frac{\pi\times 2{\cdot}35^2}{4}\times 1025=4450$ kg

Strength in bearing=23·5×1×2360+5560 kg
Least strength of rivet=4450 kg
Try 3—22 mm ϕ

Direct force on the rivet $=\frac{F}{3}=\frac{4620}{3}=1540$ kg

Max stress in the rivet due to moment $=\frac{Mr}{\Sigma r^2}$

$=\frac{75000\times 12{\cdot}2}{2\times 12{\cdot}2^2}=3060$ kg

Resultant stress in the rivet $=\sqrt{1540^2+3060^2}=3370$ kg
$\nless$ 4450 kg

1·7 Column Base

Area of base plate $=\frac{P}{\text{Permissible stress in comp in concrete}}$

$=\frac{200}{400}=0{\cdot}5\ \text{m}^2$

Try a square base plate, side of the plate $=\sqrt{\cdot 5}=0{\cdot}706$ m
$=70{\cdot}6$ cm

Use 75×75 cm on base plate

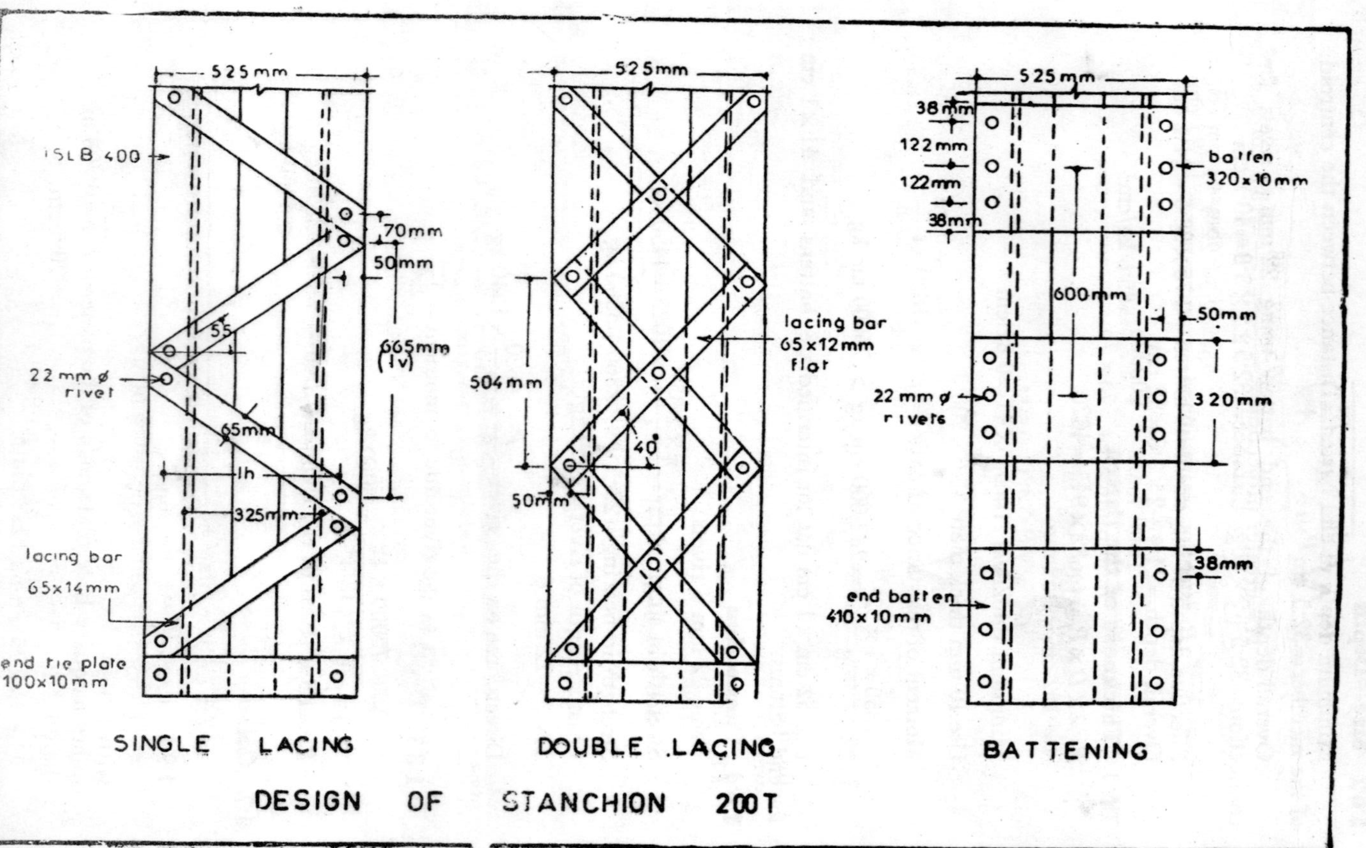

525 mm
ISLB 400
70mm
50mm
55
665mm (lv)
22 mm ø rivet
65mm
lh
325mm
lacing bar 65x14mm
end tie plate 100x10mm
SINGLE LACING
525mm
504 mm
40°
50mm
lacing bar 65x12mm flat
DOUBLE LACING
525 mm
38mm
122mm
122mm
38mm
batten 320x10 mm
600mm
50mm
22 mm ø rivets
320 mm
38mm
end batten 410x10 mm
BATTENING
DESIGN OF STANCHION 200T

1·7·1 Thickness

$$t=\sqrt{\frac{3w}{P_{bct}}\left(A^2-\frac{B^2}{4}\right)}$$

$$A=\text{greater projection}=\frac{75-42}{2}=\frac{33}{2}=15{\cdot}5 \text{ cm}$$

$$B=\text{smaller projection}=\frac{75-52{\cdot}5}{2}=\frac{22{\cdot}5}{2}=11{\cdot}25 \text{ cm}$$

$$w=\frac{\text{Total load}}{\text{Base plate area}}=\frac{200000}{75\times75}=35{\cdot}6 \text{ kg/cm}^2$$

$$P_{bct}=1890 \text{ kg/cm}^2$$

$$t=\sqrt{\frac{3\times35{\cdot}6}{1890}\left(15{\cdot}5^2-\frac{11{\cdot}25^2}{4}\right)}$$

$$=\sqrt{\frac{3\times35{\cdot}6\times208{\cdot}3}{1890}}=\sqrt{11{\cdot}8}=3{\cdot}44 \text{ cm}$$

Use 750×750×35 mm base plate.

1·7·2 Cleat Angle

Use ISA 100×100×10 mm angles with 4—22 mm ϕ rivets on the flange sides and ISA 75×75×8 mm with 3—22 mm ϕ in the webs as dummy joints to prevent any lateral movement.

1·8 Concrete Slab

Load=200 tonnes.

Self weight @ 10%=20 tonnes

Total load on soil=220 tonnes.

$$\text{Area of concrete slab}=\frac{\text{load}}{\text{BC of soil}}=\frac{220}{30}=7{\cdot}33 \text{ sqm}$$

Use square block

Side of block= $\sqrt{7{\cdot}33}=2{\cdot}72$ m

Use=2·75 m×2·75 m square

1·8·1 Depth

Assuming 45° load dispersion

$$\text{Depth of block}=\tfrac{1}{2}\,(275-75)=\frac{200}{2}=100 \text{ cm}$$

Use 2·75×2·75×1 m concrete block

1·9 Gussetted Base

$$\text{Area of base plate}=\frac{200}{400}=0{\cdot}5 \text{ m}^2$$

1·9·1 Base Plate Size

Using ISA 150×100×12 mm gusset angles on the flange sides with 100 mm leg horizontal with a gusset plate 12 mm thick minimum length required allowing 1·5 cm projection on either side in the direction parallel to webs.

$$=42+10+10+3+1{\cdot}2+1{\cdot}2=67{\cdot}4 \text{ cm}$$

Length of base plate

$$=\frac{0\cdot5\times100\times100}{67\cdot4}=74 \text{ cm}$$

Use 75 cm × 67·5 cm plate

1·9·2 Thickness

Intensity of pressure below the plate

$$=\frac{200000}{75\times67\cdot5}=39\cdot5 \text{ kg/cm}^2$$

Cantilever over hang of the base plate

$$=\frac{67\cdot5-(42+1\cdot2+1\cdot2+1\cdot2+1\cdot2)}{2}$$

$$=\frac{67\cdot5-46\cdot8}{2}=\frac{20\cdot7}{2}=10\cdot35 \text{ cm}$$

Maximum BM on one cm strip at section XX

$$M=\frac{wl^2}{2}=\frac{39\cdot5\times10\cdot35^2}{2}=2120 \text{ cm kg}$$

Resisting moment of the plate $=\frac{1}{6}\times1\times t^2\times P_{bct}$

$$=\frac{t^2}{6}\times1650=M=2120$$

$$t^2=\frac{2120\times6}{1650}=7\cdot7$$

$t=2\cdot77$ cm, Say 28 mm

Thickness of base plate $=t-$thickness of angle leg
$=28-12=16$ mm

Use 750 × 675 × 16 mm base plate

1·9·3 Connection

For complete bearing the column in flush and the load on the connection is taken in the ratio of the outstand to the full width of the base plate.

Outstand on each side $=\frac{67\cdot5-40}{2}=\frac{27\cdot5}{2}=13\cdot75$ cm

Total outstand $=27\cdot5$ cm.

Load on the concrete $=\frac{27\cdot5}{67\cdot5}\times200=81\cdot5$ tonnes.

load on each connection $=\frac{81\cdot5}{2}=40\cdot75$ tonnes.

Using 22 mm ϕ rivets

strength in single shear $=\frac{\pi}{4}(2\cdot35)^2\times1025=4450$ kg

Strength in bearing $=2\cdot35\times1\cdot2\times2360=6650$ kg

Number of rivets $=\frac{40750}{4450}=9\cdot15$ Say 10 Nos

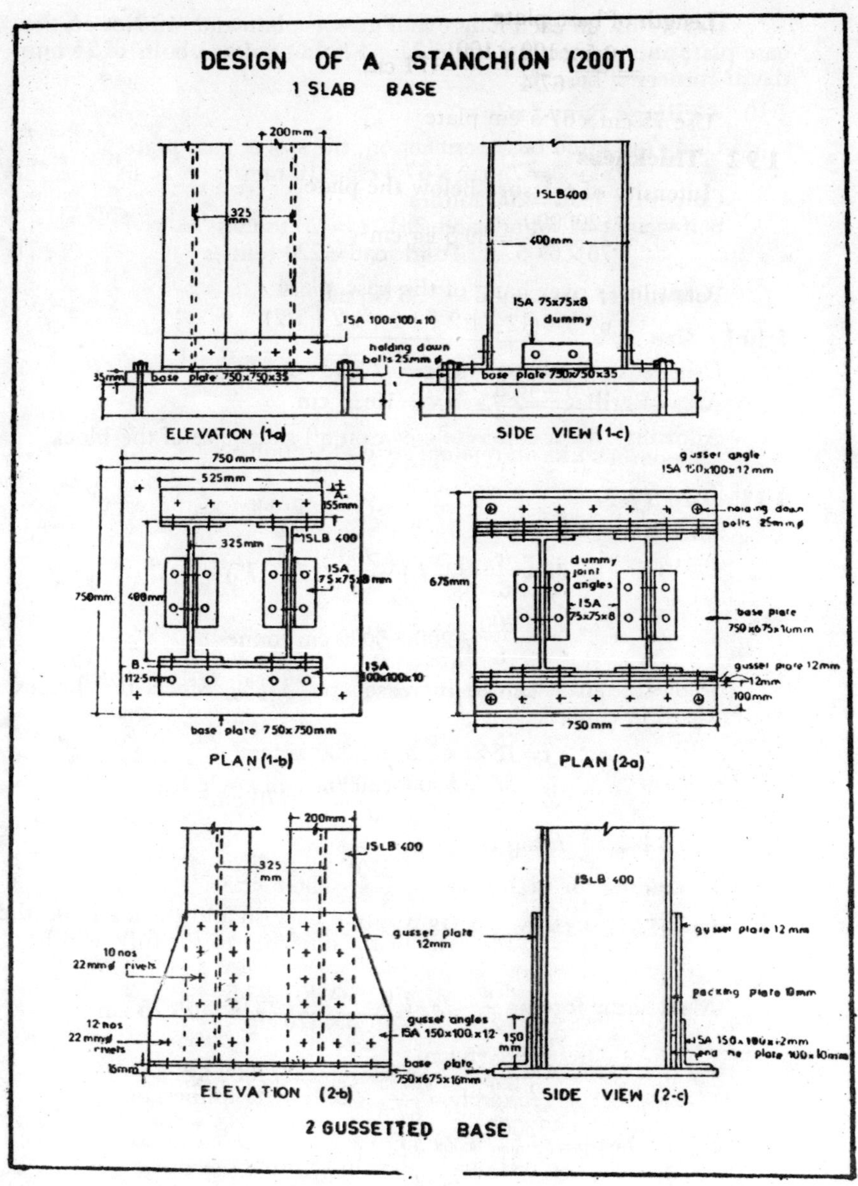
DESIGN OF A STANCHION (200T)
1 SLAB BASE
200mm
325 mm
ISA 100x100x10
holding down bolts 25mm ϕ
35mm
base plate 750x750x35
ISLB400
400mm
ISA 75x75x8 dummy
base plate 750x750x35
ELEVATION (1-a)
SIDE VIEW (1-c)
750 mm
525mm
155mm
ISLB 400
325mm
ISA 75x75x8 mm
750mm
400mm
112·5mm
ISA 100x100x10
base plate 750x750mm
PLAN (1-b)
gusset angle ISA 150x100x12 mm
holding down bolts 25mm ϕ
dummy joint angles
ISA 75x75x8
675mm
base plate 750x675x16mm
gusset plate 12mm
12mm
100mm
750 mm
PLAN (2-a)
200mm
ISLB 400
325 mm
gusset plate 12mm
10 nos 22mm ϕ rivets
12 nos 22mm ϕ rivets
gusset angles ISA 150x100x12
16mm
base plate 750x675x16mm
ISLB 400
gusset plate 12 mm
packing plate 10mm
150 mm
ELEVATION (2-b)
SIDE VIEW (2-c)
2 GUSSETTED BASE

Use 5 Nos. on each flange and gusset plate and 12 Nos on the base plate and connecting angle. Use 4 holding down bolts of 25 mm dia at corners of base.

1·10 Grillage Foundation

Using gussetted base connection, the size of base plate

$$=75\text{ cm}\times 67{\cdot}5\text{ cm}\times 16\text{ mm}$$

Column load $=200$ tonnes

Self weight of foundation @ 10% $=20$ tonnes

Total load $=220$ tonnes

Bottom tier area $=\dfrac{220}{30}=7{\cdot}33$ Sq m

1·10·1 Size

Using square grillage, side $=\sqrt{7{\cdot}33}=2{\cdot}72$ m

Area of grillage $=2{\cdot}75\text{ m}\times 2{\cdot}75\text{ m}$

Allowing 15 cm concrete cover on all sides, size of the block

$$=2{\cdot}9\text{ m}\times 2{\cdot}9\text{ m}$$

1·11 Top Tier

Maximum BM

$$M=\frac{w}{8}(L-l_1)=\frac{200}{8}(275-75)$$

$$=\frac{200}{8}\times 200=5000\text{ cm tonnes}$$

Allowable stress can be increased by $33\frac{1}{3}\%$, since the beams are encased in concrete

$$f=1650\times 1{\cdot}33=2200\text{ kg/cm}^2$$

$$Z=\frac{M}{f}=\frac{5000\times 1000}{2200}=2270\text{ cm}^3$$

Try 3 beams in top tier

Z required for each beam $=\dfrac{2270}{3}=756{\cdot}6\text{ cm}^3$

Use ISMB$-$350 $Z_{xx}=778{\cdot}9\text{ cm}^3$

1·11·1 Max. SF

Max. shear force $=\dfrac{w}{2L}(L-l_1)=\dfrac{200}{2\times 275}(275-75)$

$$=\frac{200\times 200}{2\times 275}=73\text{ tonnes}=73000\text{ kg}$$

SF per beam $=\dfrac{73000}{3}=24{,}333$ kg

Average shear stress $=\dfrac{24333}{t_w\times h}=\dfrac{24333}{35\times 0{\cdot}89}=780\text{ kg/cm}^2$

$$<945\text{ kg/cm}^2$$

Distance between flanges $=\dfrac{1}{2}(67{\cdot}5-3\times 14)=12{\cdot}75$ cm

$\not<7{\cdot}5$ cm (minimum gap between the two beams).

1·11·2 Web Buckling

$$\text{Slenderness ratio} = \frac{l}{r} = \frac{h_1\sqrt{3}}{t_w} = \frac{(35-2\times3{\cdot}1)\times1{\cdot}732}{0{\cdot}89} = 56$$

$$\text{Allowable stress } p_c = 1172-(1172-1130)\,\frac{6}{10} = 1146{\cdot}8 \text{ kg/cm}^2$$

Safe load that the beams can carry $= 1146{\cdot}8\times110\times0{\cdot}89\times3$
$= 339$ tonnes > 200 tonnes

Therefore use 3—ISMB 350 @ 52·4 kg/m for top tier.

1·11·3 Stiffeners

Provide bearing stiffeners ISA 50×50×8 mm two numbers per beam. Use 16 mm ϕ to connect them to the web on either side.

1·11·4 Separators

To keep the beam in their position, use pipe separators made of 12 mm G.I. pipe and 10 mm dia bolts.

1·11·5 Bottom Tier BM

$$M = \frac{w}{8}\,(L-l_2) = \frac{200}{8}\,(275-67{\cdot}5) = \frac{200}{8}\times207{\cdot}5$$
$$= 5200 \text{ cm tonnes.}$$

$$Z \text{ required} = \frac{5200\times1000}{1650\times1{\cdot}33} = 2360 \text{ cm}^3$$

Try 8 beams in the bottom tier

$$Z/\text{beam} = \frac{2360}{8} = 295 \text{ cm}^3$$

ISLB 250 provides $Z_{xx} = 297{\cdot}5$ cm

$$\text{Spacing} \quad = \frac{1}{7}\,(275-12{\cdot}5\times8) = 26{\cdot}6 \text{ cm} > 7{\cdot}5 \text{ cm}$$

Use 8 beams of ISLB 250.

1·11·6 Shear Force

$$\text{Maximum } SF = \frac{w}{2L}\,(L-l_2) = \frac{200}{2\times75}\,(275-67{\cdot}5)$$

$$= \frac{200}{2\times275}\times207{\cdot}5 = 75{\cdot}2 \text{ tonnes} = 75200 \text{ kg}$$

$$SF/\text{beam} = \frac{75200}{8} = 9450 \text{ kg}$$

$$\text{Average shear stress} = \frac{9450}{t_w\times d} = \frac{9450}{25\times0{\cdot}61}$$
$$= 620 \text{ kg/cm}^2 < 945 \text{ kg/cm}^2$$

1·11·7 Separators

ISA 50×50×6 mm—2·75 m long may be welded or bolted with 12 mm bolts to the top flanges of the lower tier at two ends.

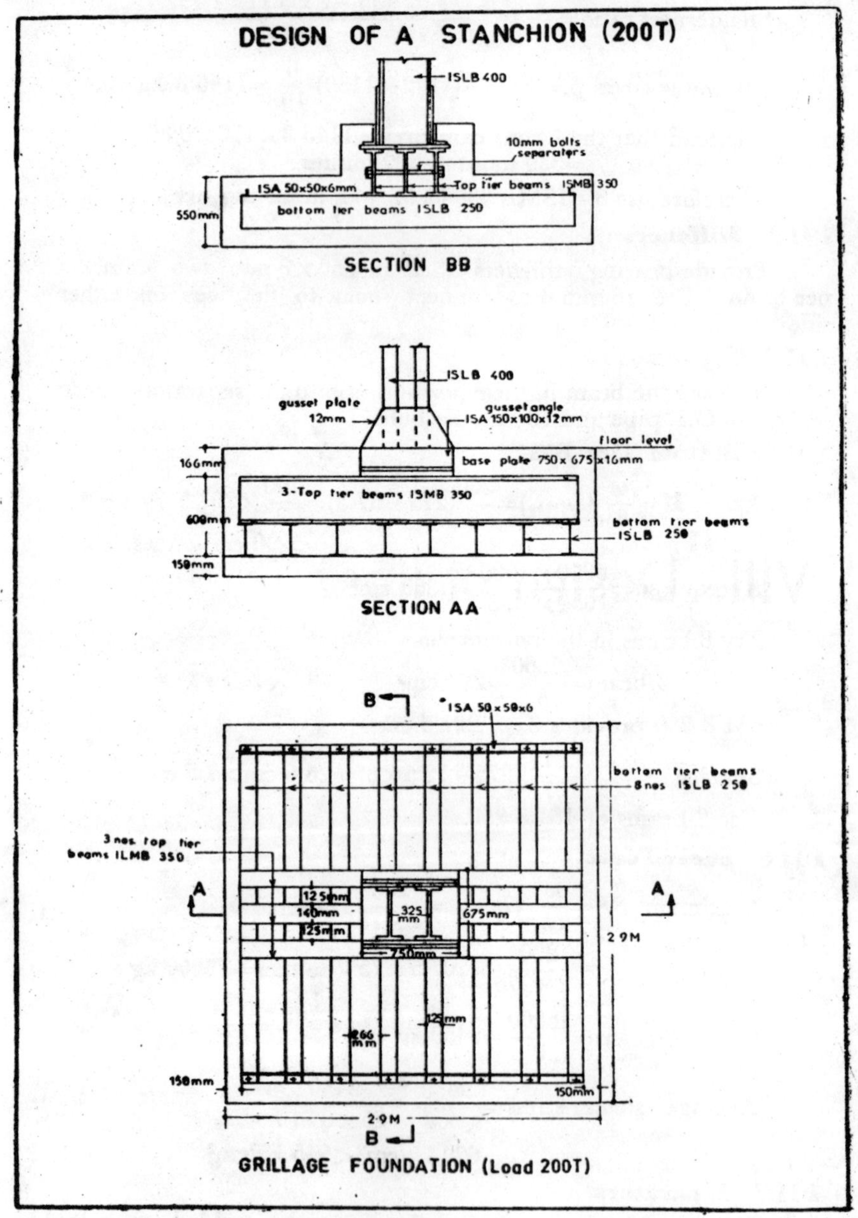
DESIGN OF A STANCHION (200T)
ISLB 400
10mm bolts separaters
ISA 50x50x6mm
Top tier beams ISMB 350
bottom tier beams ISLB 250
550mm
SECTION BB
ISLB 400
gusset plate 12mm
gusset angle ISA 150x100x12mm
floor level
166mm
base plate 750x675x16mm
3-Top tier beams ISMB 350
600mm
bottom tier beams ISLB 250
150mm
SECTION AA
B
ISA 50x50x6
bottom tier beams 8nos ISLB 250
3 nos. top tier beams ILMB 350
A
A
125mm
140mm
125mm
325 mm
675mm
750mm
2.9M
125mm
266 mm
150mm
150mm
2.9M
B
GRILLAGE FOUNDATION (Load 200T)

VIII. Design of Roof Trusses

1

Roof Trusses—General

1·1 General Information

Pitch

The pitch of the roof truss is the ratio of the height of the truss to the span. The slope of the truss depends on the type of sheeting.

Sheeting	Slope
Corrugated Iron	$26\frac{1}{2}°$ ($\frac{1}{4}$)
Asbestos sheets	20°
Slates	35°
Tiles	40°

1·2 Choice of Types

When the slope is greater than $\frac{1}{5}$ of span. Fink type trusses are used. Different types of trusses normally used for various spans are shown in Figure.

1·3 Spacing of Trusses

Common spacing of trusses ranges from 3 to 5 m. Economical spacing varies from $\frac{1}{3}$ to $\frac{1}{5}$ of span.

1·4 Spacing of Purlins

Purlins are spaced such that they are at each node of the truss to avoid bending in the main rafter of the truss. The spacing of the purlins also depends on the safe span of the sheeting material.

1·5 Covering Material

G.I. Sheets

Lengths vary from 1·2 m to 3 m with 15 cm increase in length. Safe span is 2·5 m, overlap is 15 cm.

A.C. Sheets

These sheets are available in lengths of 1·52, 1·83, 2·13, 2·74, and 3·05 m. The safe span is 1·68 m. Overlaps used 15 cm.

1·6 Loads

Dead Loads

(*a*) Sheeting

G.I. Sheeting	15 kg/m²
A.C. Sheeting	18 kg/m²
Glazing 6 mm thick	25 to 30 kg/m²

(*b*) Purlins

Self weight of purlins with corrugated sheets varies from 6 to 9 kg/m² area covered by purlin

(*c*) Trusses

$$\text{weight of truss} = W = \frac{L}{3} + 5 \text{ kg}$$

where L=span of truss

W=weight of truss/m² in kg for a spacing of 4 m for other spacing of trusses, proportional values may be taken

Weight of Wind bracing=1·5 kg/m² of plan area.

1·6·1 Live Loads

For sloping roofs of greater than 10° slope the live load is 75 kg/m² less one kg/m² for every degree increase in slope upto 20° and less 2 kg/m² for every degree increase in slope over 20° with a minimum of 40 kg/m².

1·6·2 Snow Loads

No snow load is considered if slope is greater than 50°. For other slopes 2·5 kg/m² per cm depth of snow may be taken.

1·6·3 Wind Loads

(*a*) External wind pressure+internal air pressure.

The wind pressure on roofs will be due to external wind pressure and internal air pressure. The external wind pressure acting on roofs will depend on the slope of the roof and internal air pressure on the degree of permeability of building.

Table—gives the total wind pressure for different permeabilities and slopes of roofs.

Internal wind pressure,

For small permeability internal pressure=0

For normal permeability internal pressure=$\pm 0{\cdot}2\, p$

For large openings internal pressure =$\pm 0{\cdot}5\, p$

where p=pressure intensity which depends on the zone in which the structure is located.

1·6·4 Design of Purlins

$$\text{depth of purlin} \not< \frac{L}{45}$$

width of purlin $\not< \frac{L}{60}$ when L is the spacing of trusses or span of purlin

$$\text{Maximum BM in purlin} = \frac{WL}{10}$$

where W=total distributed load on the purlin due to all loads.

1·6·5 Tension Members

(*i*) Double angle sections shall be used for main tie and single angle sections for other members.

(*ii*) Minimum size of the angle for main tie is ISA
50×50×6 mm

(*iii*) Minimum size for all other members is ISA 50×50×5 mm.

(*iv*) Minimum size of gusset plates 6 mm.

(*v*) Minimum size of rivets 16 mm.

(*vi*) Maximum permissible slenderness ratio of tension members when subjected to reversal of stress=350.

1·6·6 Compression Members

(*i*) Double angle sections are used for main rafter and for other compression members single angle sections may be used. When double angles are used for a member, they shall be unequal angles with longer legs outstanding and when single angles are used they shall be equal angles.

(*ii*) Minimum size of the angle is 50×50×6 mm for main rafter and for other compression members is ISA 50×50×5 mm.

Minimum thickness of gusset plates shall be 6 mm

Minimum size of rivets=16 mm

Minimum number of rivets at ends=2

Maximum slenderness ratio of member is 180

Effective length of member 0·7 to 1·0 of the actual length.

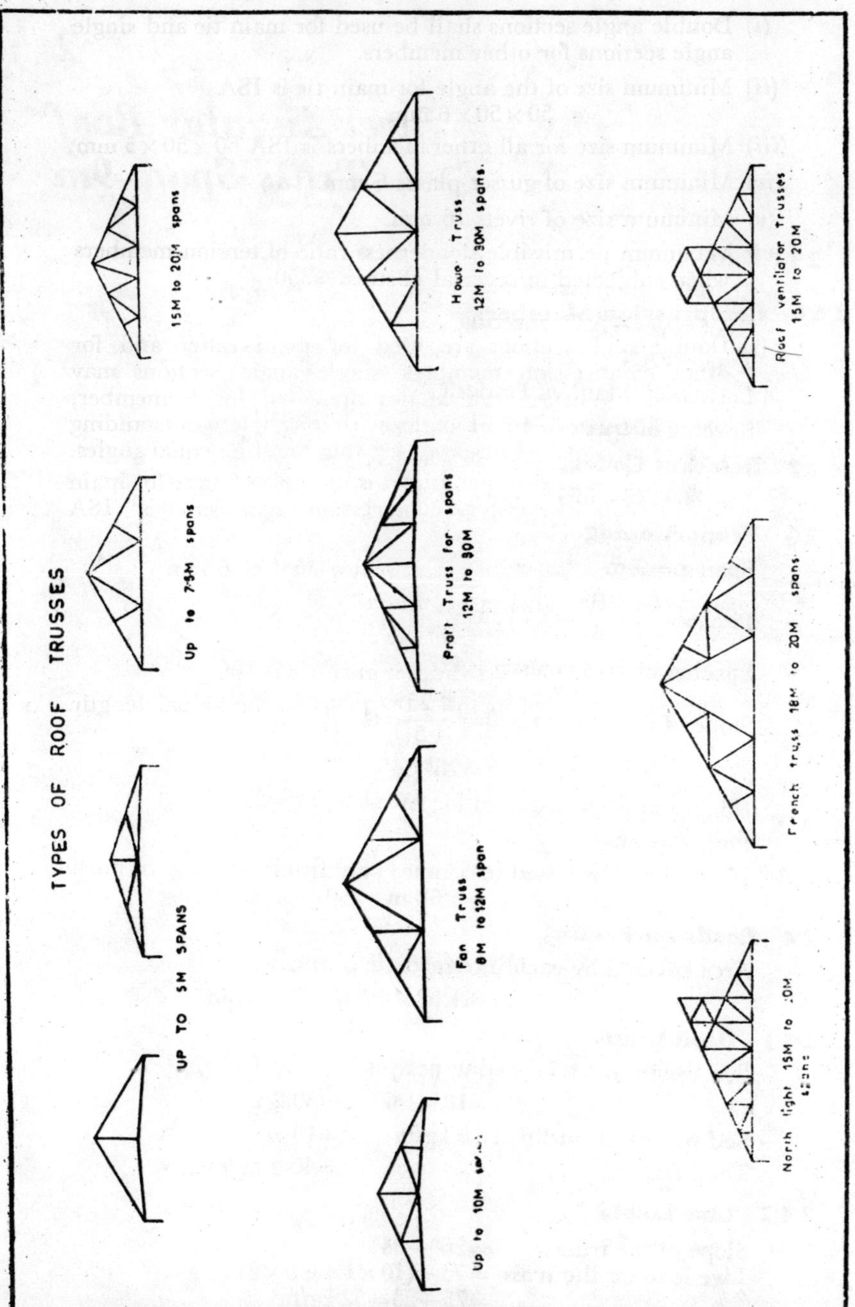
TYPES OF ROOF TRUSSES
UP TO 5M SPANS
Up to 7.5M spans
15M to 20M spans
Up to 10M spans
Fan Truss
8M to 12M span
Pratt Truss for spans
12M to 30M
Howe Truss
12M to 30M spans.
North light 15M to 20M
spans
French truss 18M to 20M spans.
Roof ventilator Trusses
15M to 20M

2

Steel Angular Roof Truss Span–9m

2·1 Data

Span for the truss=9 m

Roof cover—A.C. sheeting

Type=FAN

Location=Madhya Pradesh

Spacing of trusses=4·5 m

2·2 Relevant Codes

IS–800, IS–875

2·3 Proportioning

Span L=9 m

$$\text{Pitch}=\frac{L}{4}=\frac{9}{4}=2{\cdot}25 \text{ m}$$

Spacing of roof trusses=4·5 m

$$\text{Slope of the truss, } \tan\theta=\frac{2{\cdot}25}{4{\cdot}5}=\tfrac{1}{2}$$

$$\theta=26°\ 34'$$

Minimum slope required for AC sheeting=20°

Purlin spacing

If AC sheets are used maximnm permissible spacing of purlins =1·68 m

2·4 Loads on Purlin

Area covered by each intermediate purlin

=1·68×4·5 m=7·56 sqm

2·4·1 Dead Loads

Self weight of AC sheeting per purlin/m at 18 kg/m²

=18×1·68 m=30·2 kg

Self weight of purlin at 10 kg/m =10 kg

Total DL =40·2 kg/m

2·4·2 Live Loads

Slope of the truss =26°–34′

Live load on the truss =75–(10×1+6·5×2)

=75–23=52 kg/m²

Live load per purlin/m $= 52 \times 1{\cdot}68 \cos 26° \, 24'$
$= 52 \times 1{\cdot}68 - 0{\cdot}8942 = 78$ kg

2·4·3 Wind Loads

Wind load for (MP area upto 30 m height) $p = 150$ kg/m²

Wind pressure	Normal to ridge—slope	26°—34′
External	**Windward**	**Leeward**
Wind pressure =	$-0{\cdot}203p$	$= -0{\cdot}5p$
=	$-0{\cdot}203 \times 150$	$= -0{\cdot}5 \times 150$
=	$-30{\cdot}2$ kg/m²	$= -75$ kg/m²
Normal permeability	Windward	Leeward
Internal pressure =	$-(0{\cdot}203p + 0{\cdot}2p)$	$= -(0{\cdot}5p + 0{\cdot}2p)$
=	$-0{\cdot}403p$	$= -0{\cdot}7p$
=	$-0{\cdot}403 \times 150$	$= -0{\cdot}7 \times 150$
=	$-60{\cdot}5$ kg/m²	$= -105$ kg/m²
Internal suction =	$(-{\cdot}203p + 0{\cdot}2p)$	$= -0{\cdot}5p + 0{\cdot}2p$
=	$-{\cdot}003p$	$= -0{\cdot}3p$
=	$-0{\cdot}45$ kg/m²	$= -{\cdot}3 \times 150$
		$= -45$ kg/m²

Maximum wind load per purlin/m

$= -105 \times 1{\cdot}68 \times \cos 26° \, 24'$

$= -105 \times 1{\cdot}68 \times 0{\cdot}8942 = -158$ kg

2·4·4 Load Combination

$$DL + LL = 40{\cdot}2 + 78 = 118{\cdot}2 \text{ kg/m}$$

$$DL + WL = 40{\cdot}2 - 158 = -117{\cdot}8 \text{ kg/m}$$

When wind load is considered the load on purlin can be decreased by 50%

$\therefore$ Design load $= 118{\cdot}2$ kg/m

2·5 Bending Moment

For a continuous purlin

$$BM = \frac{WL}{10} = \frac{(118{\cdot}2 \times 4{\cdot}5)}{10} 4{\cdot}5 = 240 \text{ mkg}$$

$$Z \text{ required} = \frac{24000}{1650} = 14{\cdot}7 \text{ cm}^2$$

$$\text{Minimum width of purlin} = \frac{L}{60} = \frac{450}{60} = 7{\cdot}5 \text{ cm}$$

$$\text{Minimum depth of purlin} = \frac{L}{60} = \frac{450}{45} = 10 \text{ cm}$$

ISA 100×75×6 mm @ 8 kg/m provide $Z_{xx} = 14{\cdot}4$ cm²

ISA 100×75×8 mm @ 10·5 kg/m $Z_{xx} = 19{\cdot}1$ cm² $> 14{\cdot}7$ cm²

2·6 Load on the Truss

Dead Loads

Sloping length of rafter $=\sqrt{2{\cdot}25^2+4{\cdot}5^2}$

$=5{\cdot}03$ m

Slope $=26°—34'$

Spacing of the trusses $=4{\cdot}5$ m c/c

Weight of sheeting on $\frac{1}{2}$ truss (plan area) @ 18 kg/m²

$=4{\cdot}5\times4{\cdot}5\times18=365$ kg

Weight of purlins (4 Nos.) @ 10 kg/m$=4\times10\times4{\cdot}5=180$ kg

Self weight of roof truss$=\frac{9}{3}+5$ kg/m²$=8$ kg/m²

Weight of $\frac{1}{2}$ roof truss $=8\times4{\cdot}5\times4{\cdot}5=162$ kg

Total DL/$\frac{1}{2}$ truss $=365+180+162=707$ kg

Dead load on intermediate panel point$=\frac{707}{3}=236$ kg

Load on end panel point $=\frac{236}{2}=118$ kg

Live Load

LL on $\frac{1}{2}$ truss $=52\times4{\cdot}5\times4{\cdot}5=1055$ kg

LL on intermediate penal point$=\frac{1055}{3}=352$ kg

LL on end panel point $=176$ kg

Wind Loads

Maximum wind load on $\frac{1}{2}$ truss on the windward side

$=-60{\cdot}5\times4{\cdot}5\times5{\cdot}03=-1440$ kg

WL/intermediate panel point$=-\frac{1440}{3}=-480$ kg

WL/end penal $=-\frac{480}{2}=-240$ kg

Maximum wind load on $\frac{1}{2}$ truss on the Leeward side

$=-105\times4{\cdot}5\times5{\cdot}03=-2505$ kg

Wind load/intermediate panel$=-\frac{2505}{3}=-835$ kg

Wind load/end panel point $=-\frac{835}{2}=-417{\cdot}5$ kg

2·6·1 Forces in Truss Members

As in Table 2·6·1.

Table 2·6·1 Forces in Members of Fan Truss (Pitch $\frac{1}{4}$)

DL=1414 kg
DL=2110 kg

WL (L)=1440 kg
WL (R)=2505 kg

Member	DL kg	LL kg	Wind load kg		Load combinations			Design load	
			Left	Right	DL+LL kg	DL+WL (L) kg	DL+WL (R) kg	Compression kg	Tension kg
1	+1400	+1980	−1500	−2640	+3380	−100	−1240	3380	1240
2	+1135	+1600	−1180	−2060	+2735	− 55	− 925	2735	925
3	+1175	+1660	−1500	−2640	+2835	−325	−1465	2835	1465
4	−500	−700	+1070	+1880	−1200	+570	+1330	1330	1200
5	+270	+380	− 580	−1010	+ 650	−310	− 740	650	740
6	+270	+380	− 580	−1010	+ 650	−310	− 740	650	740
7	−950	−1760	+1890	+3300	−2710	+940	+2250	2250	2710
8	−750	−1056	+ 810	+1410	−1806	+ 60	+ 660	660	1806

2·6·2 Design—Truss Members—Members 1 to 3

Main rafter—Member is in compression

Maximum Compressive force = 3380 kg
Tension = 1465 kg
Length of member = 167 cm
Effective length = 0·7 × 167 = 117 cm

Try 2 angles ISA 50 × 50 × 6 mm back to back

(minimum size required) A = 2 × 5·68 = 11·36 cm²

$$r_{min} = 1{\cdot}51 \text{ cm}$$

$$l/r = \frac{117}{1{\cdot}51} = 77{\cdot}5 < 180$$

for $l/r = 77{\cdot}5$, $p_c = 1025{\cdot}5$ kg/cm²

When wind loads are considered this can be increased by $33\frac{1}{3}$%

Safe load = 1025·5 × 1·33 × 11·36 = 15610 kg > 3380 kg

2·6·2 Member 4

Member is in tension

Maximum tension = 1200 kg
Compression = 1330 kg
Length of member = 281 cm

Try one angle ISA 50 × 50 × 5 mm

A = 4·79 cm², $r_{min} = 1{\cdot}52$ cm

Effective area $= a + kb$

Using 10 mm ϕ rivets,

Area of connecting leg = (5 − 1·75) 0·5

$$a = 3{\cdot}25 \times 0{\cdot}5 = 1{\cdot}625 \text{ cm}^2$$

$$b = (5 - 0{\cdot}5) + 0{\cdot}5 = 4{\cdot}5 \times 0{\cdot}5 = 2{\cdot}25 \text{ cm}^2$$

$$k = \frac{1}{1 + 0{\cdot}35\dfrac{b}{a}} = \frac{1}{1 + 0{\cdot}35\dfrac{2{\cdot}25}{1{\cdot}625}} = 0{\cdot}675$$

Effective area $= a + kb = 1{\cdot}625 + {\cdot}675 \times 2{\cdot}25 = 3{\cdot}135$ cm²

Safe load in tension = 3·135 × 1500 × 1·33 = 6250 kg > 1200 kg

Check for compression

Effective length = 281 × 0·7 = 197 cm

$$r_{min} = 0{\cdot}97 \text{ cm}$$

$$l/r = \frac{197}{0{\cdot}97} = 204 < 350$$

$$\text{for } \frac{l}{r} = 204, \quad p_c = 258 \text{ kg/cm}^2$$

Safe load in compression = 4·79 × 258 × 1·33
= 1640 kg > 1330 kg

2·6·4 Member 5 & 6

Members in compression

Compressive force = 650 kg

Tension = 740 kg

Length of member = 151 cm

Effective length = 151 × 0·7 = 105·7 cm

Try minimum size of angle ISA 50 × 50 × 5 mm

$A = 4{\cdot}79\ \text{cm}^2, \quad r_{min} = 0{\cdot}97\ \text{cm}$

$l/r = \dfrac{105{\cdot}7}{0{\cdot}97} = 108{\cdot}2$ p_c for l/r 108·2 = 811 kg/cm²

Safe load in compression = 811 × 1·33 × 4·79 = 5180 kg > 650 kg

Effective tensile area for ISA 50 × 50 × 5, using 16 mm rivet = 3·135 cm²

Safe load in tension = 3·135 × 1500 × 1·33 = 6250 kg > 740 kg

2·6·5 Members 7 and 8 (Main Tie Member)

Maximum tensile force = 2710 kg

Compressive force = 2250 kg

Length of member = 281 cm

Try minimum two angles back to back, ISA 50 × 50 × 6

Gross area provided A = 2 × 5·68 = 11·36 cm²

Using 16 mm ϕ rivets

Net area = 11·36 − 2 × 1·75 × 0·6 = 9·26 cm²

Safe load in tension = 9·26 × 1500 × 1·33 = 18500 kg > 2710 kg

Effective length = 0·7 × 281 = 196·7 cm

$r_{min} = 1{\cdot}51\ \text{cm}$

$l/r = \dfrac{196{\cdot}7}{1{\cdot}51} = 131 < 350$

for $l/r = 131, \quad p_c = 622\ \text{kg/cm}^2$

Safe load in compression = 622 × 1·33 × 11·36
= 9350 kg > 2250 kg

2·7 Design of Joints—Joint A

Member 1 and 7 meet at the point

Maximum force in 1 = 3380 kg

Maximum force in 7 = 2710 kg

Using 16 mm ϕ rivets with 6 mm thick guesset plates strength in single shear $= \dfrac{\pi \times 1{\cdot}75^2}{4} \times 1025 = 2465$ kg

In Double shear = 4930 kg

Strength in bearing on 6 mm thick plate

$$=1{\cdot}75\times0{\cdot}6\times2360=2478 \text{ kg}$$

Rivets required in member 1 (rivets are in double shear, therefore the least strength is in bearing) $=\frac{3380}{2478}\simeq2$

Rivets required in member 7 $=\frac{2710}{2478}\simeq2$

2·7·1 Joint B

Members 1, 2 and 6 meet at the joint

Force in 1 = 3380 kg

Force in 2 = 3380 kg

Force in 6 = 650 kg

Rivets are in double shear and therefore the least strength of rivet is in bearing = 2478 kg

Members 1 and 2 are continuous

The force in the direction of this member due to component of force in member 6 is $=650 \cos(90°-26° \,34')=650 \cos 63° \,26'$

$$=650\times{\cdot}4473=292 \text{ kg}$$

Number of rivets $=\frac{292}{2478}\simeq1$

However use minimum of three rivets

Number of rivets in Member 6 $=\frac{650}{2465}\simeq1$

(rivets are in single shear)

Use a minimum of 2 rivets

2·7·2 Joint C

Members 2, 3 and 5 meet at this joint

Members 2 and 3 are continuous. Use minimum of three rivets

Number of rivets in member 5 $=\frac{650}{2465}\simeq1$

Use a minimum of two numbers

2·7·3 Joint D

Member 3, 4 and 3′, 4′ meet at this joint

Force in members 3 and 3′ = 2835 kg

Force in 4 and 4′ = 1330 kg.

Number of rivets required for 3 and 3′ $=\frac{2835}{2478}\simeq2$

Number of rivets required in 4 and 4′ $=\frac{1330}{2465}\simeq1$

Use a minimum of two rivets.

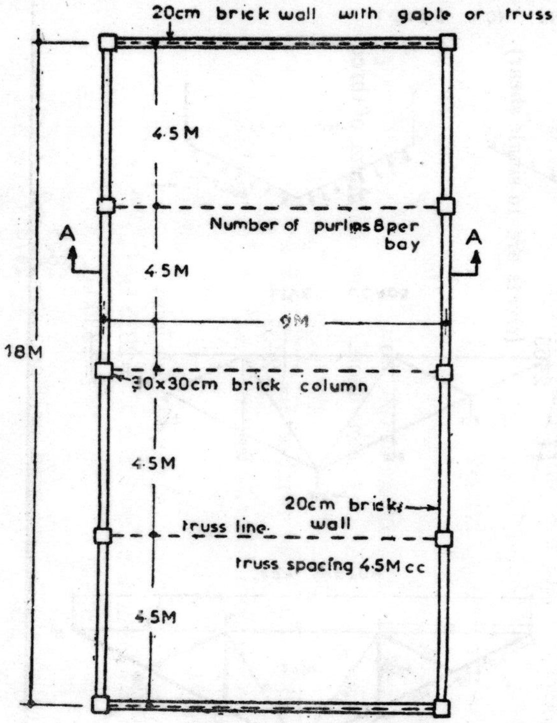

20cm brick wall with gable or truss

4·5M

Number of purlins 8 per bay

A

A

4·5M

9M

18M

30x30cm brick column

4·5M

20cm brick wall

truss line

truss spacing 4·5M cc

4·5M

PLAN SHED SIZE 18M x 9M

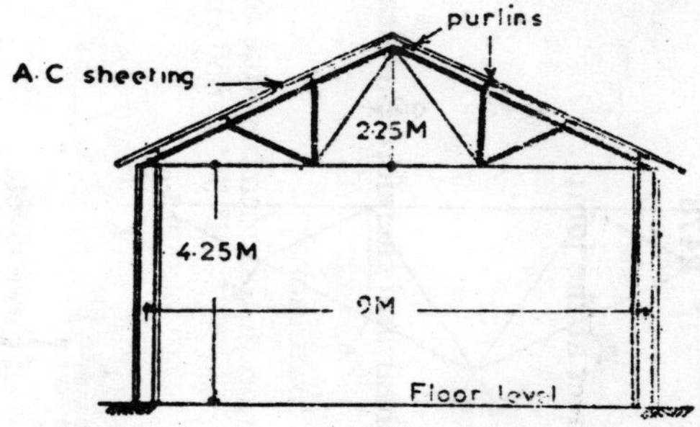

purlins

A·C sheeting

2·25M

4·25M

9M

Floor level

SECTION AA

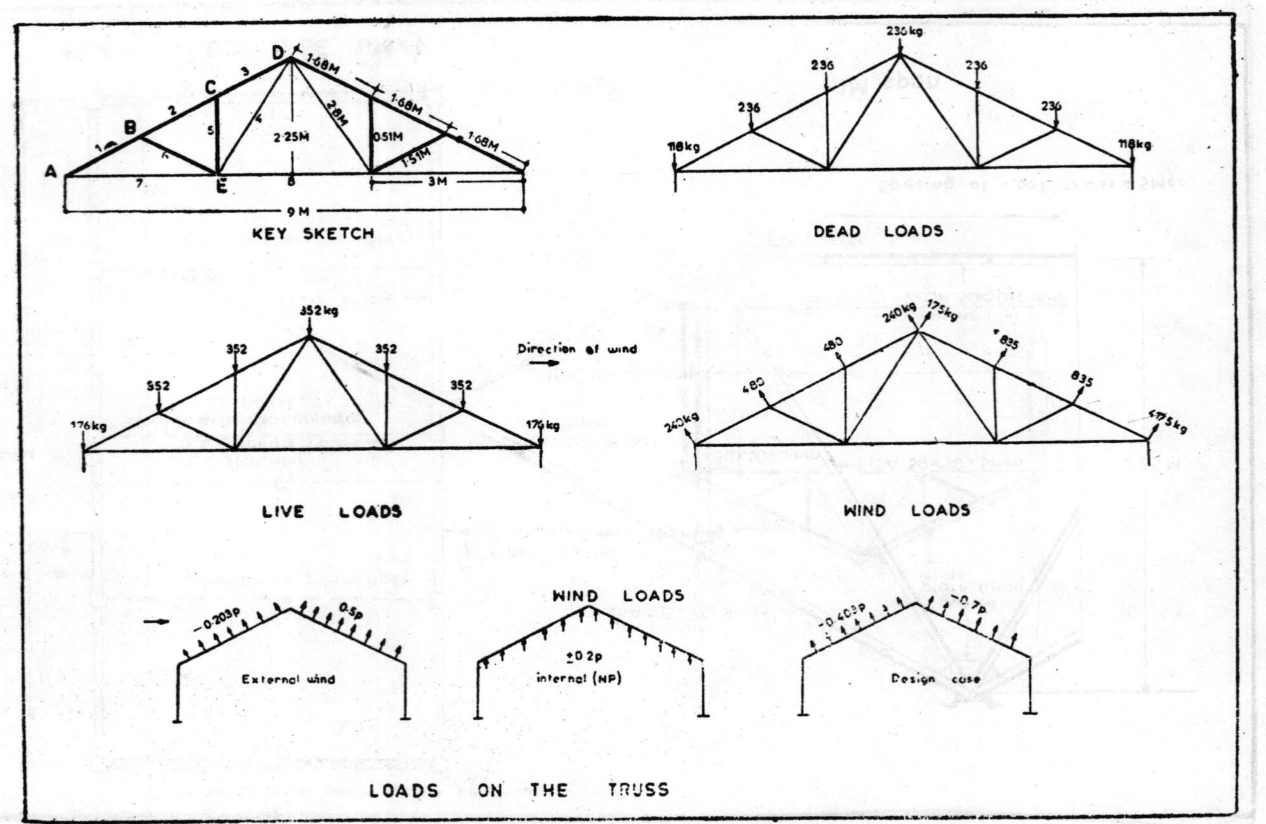

LOADS ON THE TRUSS

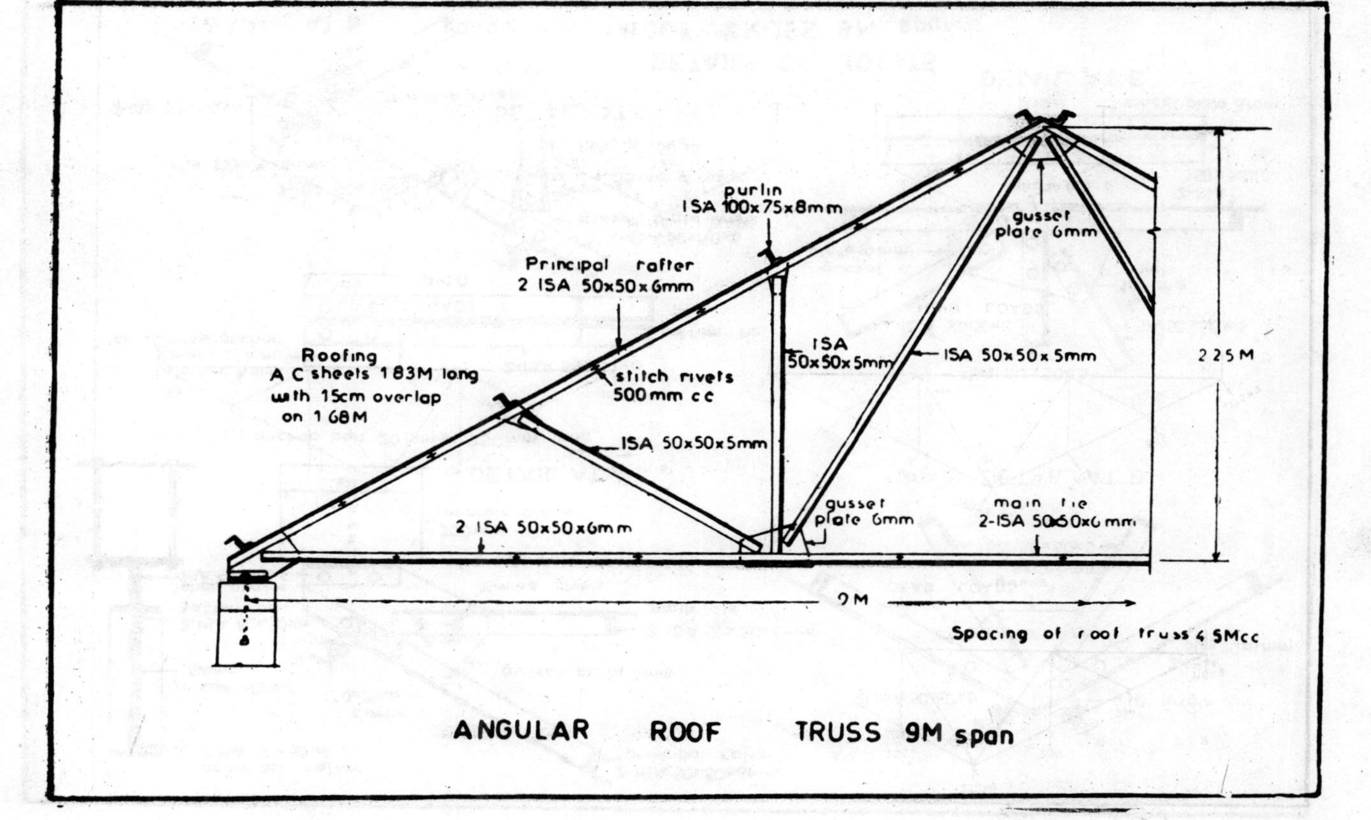

ANGULAR ROOF TRUSS 9M span

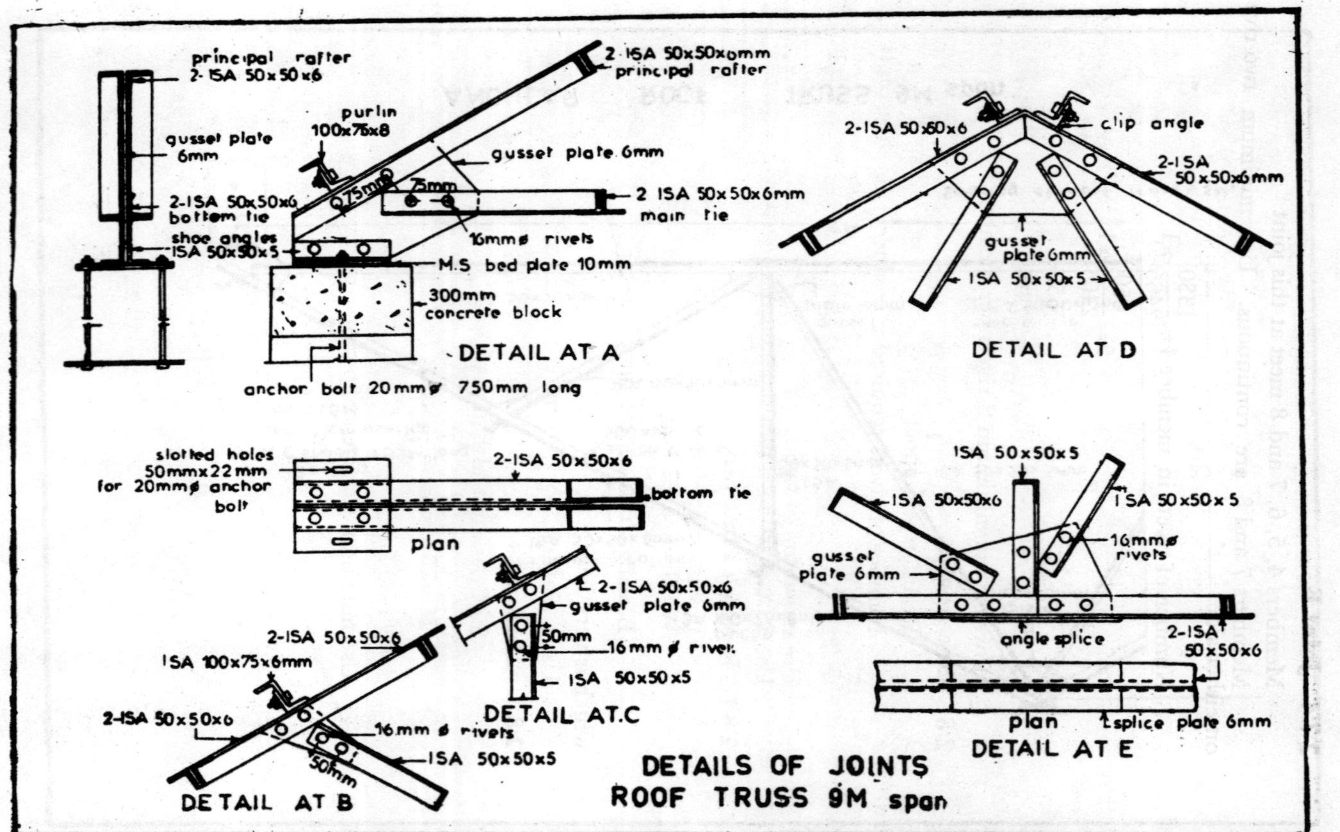

DETAILS OF JOINTS
ROOF TRUSS 9M span

2·7·4 Joint E

Members 4, 5, 6, 7 and 8 meet at this joint

Members 7 and 8 are continuous. Use minimum two rivets on either side

Number of rivets in member $4=\frac{1330}{2465}\simeq 1$

,, ,, $5=\frac{650}{2465}\simeq 1$

,, ,, $6=\frac{650}{2465}\simeq 1$

However use minimum 2 rivets per member

2·8 Shoe Angles

Reaction at support $=1762$ kg

Number of rivets required to connect the shoe angle

$$=\frac{1762}{2473}\simeq 1$$

Use minimum of two rivets

Use ISA 75×75×6 mm shoe angle 30 cm long.

2·8·1 Bearing Plate Size

Bearing plate length $=30$ cm

Width$=7{\cdot}5\times 2+0{\cdot}6$ $=15{\cdot}6$ cm

Bearing stress on masonry$=\frac{1762}{30\times 15{\cdot}6}=3{\cdot}72$ kg/cm^2

which is very low

Use a plate of 30×15·6 cm

2·8·2 Thickness

BM per cm width of plate$=3{\cdot}72\times 7{\cdot}5\times 1$

$$M=28 \text{ cm kg}=\frac{1}{6}\,bt^2\times f=1800\times\frac{1}{6}\times 1\times t^2$$

$t=0{\cdot}3$ cm

Use minimum thickness of 6 mm for bearing plate.

3

Tubular Steel Truss Span—18 m

3·1 Data

Size of the shed $=18 \text{ m} \times 27 \text{ m}$

Span for truss $=18$ m

Location—Madhya Pradesh

3·2 Relevant Codes

IS—800 : Code of practice for use of structural steel in general building construction.

IS—806 : Use of steel tubes in general building construction.

IS—875 : Code of practice for structural safety of buildings—Loading standards.

IS—1161 : Specification for steel tubes for structural purposes.

IS—816 : Code of practice for use of metal arc welding for general construction in mild steel.

IS—814 : Specification for covered electrodes for metal arc welding of mild steel.

IS—226 : Specification for structural steel

8·3 Truss Proportioning

Span $=18$ m

Pitch for the truss $=\dfrac{L}{4}=\dfrac{18}{4}=4\cdot5$ m

Type=French or compound Fink truss.

Spacing of the trusses $=4\cdot5$ m

Slope of truss $=\tan\theta=\dfrac{4\cdot5}{9}=\dfrac{1}{2}=26^\circ\,34'$

Minimum slope required for AC sheeting $=20^\circ$

Therefore pitch used is sufficient for weather tight joints.

Purlin spacing

If AC sheets are used maximum permissible spacing of purlins $=1\cdot68$ m

If C. G. sheets are used maximum permissible spacing is 2·75 m

3·4 Purlin Loads

Use AC sheets

Dead loads

Area covered by each purlin $=1{\cdot}68\times4{\cdot}5$ m

Self wt of AC sheeting per purlin/m $=18\times1{\cdot}68=30{\cdot}2$ kg

Self wt of purlin at 10 kg/m $=10{\cdot}0$ kg

Total DL/m $=40{\cdot}2$ kg

Live Loads

Slope of the truss $=26°\ 34'$

Live load on the truss $=75-(10\times1+6{\cdot}5\times2)=75-23$

$=52$ kg/m²

Live load per purlin/m $=52\times1{\cdot}68\cos 26°\ 34'$

$=52\times1{\cdot}68\times0{\cdot}8942=78$ kg

Wind loads

Wind load for Madhya Pradesh area up to 30 m height $=p=150$ kg/m². Wind pressure normal to ridge.

Slope 26° 34′

	Windward	Leeward
External Wind pressure	$=-0{\cdot}203\,p$	$=-0{\cdot}5\,p$
	$=-0{\cdot}203\times150$	$=-0{\cdot}5\times150$
	$=-30{\cdot}2$ kg/m²	$=-75$ kg/m²

Normal permeability	Windward	Leeward
Internal pressure	$=-({\cdot}203\,p+0{\cdot}2\,p)$	$=-(0{\cdot}5+{\cdot}2\,p)$
	$=-{\cdot}403\,p$	$=-0{\cdot}7\,p$
	$=-{\cdot}403\times150$	$=-0{\cdot}7\times150$
	$=-60{\cdot}5$ kg/m²	$=-105$ kg/m²
Internal Suction	$=(-0{\cdot}203+0{\cdot}2\,p)$	$=-{\cdot}5\,p+0{\cdot}2\,p$
	$=-{\cdot}003\,p$	$=-0{\cdot}1\,p$
	$=-{\cdot}003\times150$	$=-0{\cdot}3\times150$
	$=-0{\cdot}45$ kg/m²	$=-45$ kg/m²

Max. wind load per purlin/m $=-105\times1{\cdot}68\times\cos 26°\ 34'$

$=-105\times1{\cdot}68\times0{\cdot}8942$

$=-158$ kg.

3·4·1 Load Combination

DL+LL $=40{\cdot}2+78=118{\cdot}2$ kg/m

DL+WL $=40{\cdot}2-158=-117{\cdot}8$ kg/m

When wind load is considered, the load on purlin can be decreased by 50%.

However the purlin has to carry more load under DL+LL combination.

Hence design load $=118{\cdot}2$ kg/m

3·4 Bending Moment

Continuous purlin

$$\text{BM}=\frac{WL}{10}=\frac{(118{\cdot}2\times4{\cdot}5)\ 4{\cdot}5}{10}=240 \text{ m kg}$$

$$Z \text{ required}=\frac{24000}{1400}=17{\cdot}1 \text{ cm}^3$$

Size of purlin as required by IS—806

$$Z \text{ required}=\frac{WL}{16800}=\frac{118{\cdot}2\times4{\cdot}5\times4{\cdot}5\times100}{16800}=14{\cdot}3 \text{ cm}^3$$

Where W=total distributed load in kg on the purlin arising from dead load and snow load but excluding wind

L=the distance in cm between the centres of the steel principals or other supports.

$W=40{\cdot}2\times4{\cdot}5$ kg

$L=450$ cm

$$Z \text{ required}=\frac{(40{\cdot}2\times4{\cdot}5)\times450}{16800}=\frac{180\times450}{16800}=\frac{81000}{16800}=4{\cdot}82 \text{ cm}^3$$

Try 40—M (48·3—OD) Z $=4{\cdot}86$ cm³

Size of purlin based on the limiting deflection

Bending stress $=31500\ \frac{D}{L}$

$$1400=\frac{31500\ D}{450}$$

$$D=\frac{450\times1400}{31500}=20 \text{ cm}^3$$

Limiting deflection need not be checked if

$$\text{Size}=\frac{L}{70}=\frac{450}{70}=6{\cdot}43 \text{ cm}$$

65 mm provided

Hence use 65 mm—M

2·5 Load On The Truss

Dead loads

Slopping iength of rafter $=\sqrt{4{\cdot}5^2+9^2}=10{\cdot}06$ m

Slope $=26° 34'$

Spacing of trusses $=4{\cdot}5$ m

Wt. of sheeting on ½ truss (plan area)

$=9\times4{\cdot}5\times18=730$ kg.

Wt of purlins (7 Nos) @ 10 kg/m $=7\times4{\cdot}5\times10=315$ kg
(@1·68 m c/c) on ½ truss

Self weight of roof truss $=\frac{18}{1}+5=11$ kg/m²

Wt. of ½ truss $=11\times4{\cdot}5\times9=446$ kg

Total DL/½ truss $=730+315+446=1491$ kg

Dead load on intermediate panel point

$=\frac{1491}{4}=373$ kg say 375 kg.

Dead load on end panel $=\frac{375}{2}=187{\cdot}5$ kg

Live loads :

L on ½ truss $=52\times4{\cdot}5\times9=2110$ kg

L. load/intermediate panel pt. $=\frac{2110}{4}=528$ kg

L.L/end panel point $=\frac{528}{2}=264$ kg

Wind loads :

Max. *WL* on ½ truss wind ward side

$=-60{\cdot}5\times4{\cdot}5\times10{\cdot}06=-2880$ kg

W. load/intermediate panel

$=-\frac{2380}{4}=-720$ kg

W. load-end panel $=360$ kg

Max. *WL* on ½ truss Leeward side

$=-105\times4{\cdot}5\times10{\cdot}06=-5010$ kg.

W. load intermediate panel

$=-\frac{5010}{4}=-1252{\cdot}5$ kg

Say $=-1255$ kg

W. load/end panel pt $=-627{\cdot}5$ kg

3·6·1 Forces In Truss Members

As in Table—3·6·1

TABLE 3·6·1 Forces In kg In The Truss Member

+Compression —Tension

Member	DL	LL	Wind load		Lord combinations			Design Loads	
			Left	Right	DL+LL	DL+WL(L)	DL+WL(R)	Max Comp.	Max. Tension
1	+2950	+4150	−6350	−7450	+7100	−3400	−4500	+7100	−4500
2	+2770	+3920	−6350	−7450	+6690	−3580	−4680	+6690	−4680
3	+2610	+3650	−6350	−7450	+6260	−3740	−4840	+6260	−4840
4	+2440	+3430	−6350	−7450	+5870	−3910	−5010	+5870	−5010
5	−750	−1030	+1620	+3240	−1780	+870	+2490	+2490	−1780
6	−1115	−1600	+2420	+4160	−2715	+1305	+3045	+3045	−2715
7	−2250	−3200	+3760	+5900	−5450	+1510	+3650	+3650	−5450
8	−2530	−3720	+4600	+7200	−6220	+2070	+4670	+4670	−6220
9	+335	+475	−720	−1250	+810	−385	−915	+ 810	− 915
10	−375	−530	+810	+1400	−905	+435	+1025	+1025	− 905
11	+675	+950	−1460	−2500	+1625	−785	+1825	+1825	− 785
12	−375	−530	+810	+1400	−905	+435	+1025	+1025	− 905
13	+335	−475	−720	−1250	+810	−385	−915	+ 810	− 915
14	−1500	−2120	+2160	+3100	−3620	+660	+1600	+1600	−3620

3·7 Design of Truss Members—Main rafter—Member 1 to 4

Members 1 to 4 (compression members)

Max. comp. force=7100 kg

Max. tensile force=5010 kg

Length of the member=251·5 cm

Try 65 mm *ID* light section *OD*=76·1 mm

$A=7{\cdot}44$ cm², $r=2{\cdot}58$ cm

$$\frac{l}{r}=\frac{251{\cdot}5\times0{\cdot}7}{2{\cdot}58}=68{\cdot}5$$

Allowable stress for

$l/r=69$, $f_c=1002$ kg/cm²

$l/r=70$, $f_c=970$ kg/cm²

By interpolation, l/r 68·5 $f_c=974{\cdot}8$ kg/cm²

safe load=$f_c\,A=974{\cdot}8\times7{\cdot}44=7300$kg > 7100 kg

Therefore use 65 mm *ID*—Light tube for main rafter.

Tensile stress $=\dfrac{5010}{7{\cdot}44}=675$ kg/cm² > 1250 kg/cm²

Hence member is safe in tension.

3·7·1 Members 5 & 6

Length of member =281 cm
Max. comp. force =2945 kg
Max. tension =2715 kg

Member is in tension under dead and live load.

Try 32 mm *ID*—Medium, *OD* 42·4 mm

$A=4{\cdot}00$ cm², $r=1{\cdot}39$ cm

Area required for tension$=\dfrac{2715}{1250}=2{\cdot}17$ cm² < 4 cm² provided

Hence safe in tension

Check for compression

Length of the member=281 cm

$$\frac{l_e}{r}=\frac{281\times0{\cdot}7}{1{\cdot}39}=142<180$$

$l/r=140$, $f_c=540$ kg/cm²

$l/r=150$, $f_c=490$ kg/cm²

$l/r=142$, $f_c=530$ kg/cm²

When wind loads are considered the stress can be increased by 33⅓%.

Allowable stress=1·33 × 530=706 kg/cm²

Safe load =706 × 4 =2824 kg < 3045 kg

Hence unsafe

Try 40 mm—*ID*—*M*, $OD=48{\cdot}3$ mm

$A=4{\cdot}60\ \text{cm}^2,\quad r=1{\cdot}60$ cm

$$\frac{l_e}{r}=\frac{281\times0{\cdot}7}{1{\cdot}60}=123<180$$

$l/r=120,\quad f_c=674\ \text{kg/cm}^2$
$l/r=130,\quad f_c=605\ \text{kg/cm}^2$
$l/r=123,\quad f_c=653{\cdot}3\ \text{kg/cm}^2$

Allowable stress$=1{\cdot}33\times653{\cdot}3=875\ \text{kg/cm}^2$
Safe load $=875\times4{\cdot}6=4000\ \text{kg}>3045$ kg
Hence use 40 mm *ID*—Medium section

3·7·2 Members 7, 8, 14—Main Bottom Tie Member

Max. comp. force $=4670$ kg

Area required for tension$=\frac{6220}{1260}=5\ \text{cm}^2$

Try 50 mm *ID* L_2 section $OD=60{\cdot}3$ cm

$A=5{\cdot}82\ \text{cm}^2,\quad r=2{\cdot}02$ cm

Hence safe in tension
Check for compression

Length of the member$=281$ cm

$$\frac{l_e}{r}=\frac{281\times0{\cdot}7}{2{\cdot}02}=97{\cdot}5$$

$l/r=90,\quad f_c=876\ \text{kg/cm}^2$
$l/r=100.\quad f_c=814\ \text{kg/cm}^2$
$l/r=97{\cdot}5,\quad f_c=829{\cdot}5\ \text{kg/cm}^2$

Allowable stress$=1{\cdot}33\times829{\cdot}5=1100\ \text{kg/cm}^2$
Safe load $=1100\times5{\cdot}82=6410\ \text{kg}>4670$ kg

Hence use 50 mm *ID*—L_2—section

3·7·3 Members 9 and 13

Members are in compression under DL+LL.
Maximum component force$=810$ kg
Maximum Tensile force$=915$ kg
Length of the member$=126$ cm
Try *ID* 20—Heavy section, $OD=26{\cdot}9$ mm.

$A=241\ \text{cm}^2,\quad r=0{\cdot}84$ cm

$$\frac{l_e}{r}=\frac{126\times0{\cdot}7}{{\cdot}84}=105$$

$l/r=100,\quad f_c=814\ \text{kg/cm}^2$
$l/r=110,\quad f_c=745\ \text{kg/cm}^2$
$l/r=105,\quad f_c=779{\cdot}5\ \text{kg/cm}^2$

Safe load$=2{\cdot}41\times779{\cdot}5=1930\ \text{kg}>810$ kg
Uneconomical

Try *ID*—15—H, $A=1{\cdot}84\ \text{cm}^2$
$r=0{\cdot}65$ cm

$$l_e/r=\frac{126\times0{\cdot}7}{0{\cdot}65}=135<180$$

Allowable stress =571·5 kg/cm².
Safe load=571·5×1·84=1050 kg > 810 kg
Hence safe.
Check for tension.

Area required for tension=$\frac{915}{1250}$=0·735 cm²
< 1·84 cm² provided.

Use ID=15 mm. Heavy section, OD=21·3 mm

3·7·4 Members 10 and 12

Members are in tension under DL+LL
Maximum tension=905 kg
Maximum comp. =1025 kg

Area required for tension=$\frac{905}{1250}$=0·745 cm²

Try ID—15 mm—H

A=1·84 cm², r=0·65 cm

Area provided is greater than required hence safe in tension.

Check for compression.

l_e=281 cm

$l_e/r=\frac{281\times0·7}{0·65}=305 < 350$

Not sufficient

Try ID=25 mm—M

A=3·11 cm², r=1·08 cm

$l_e/r=\frac{281\times0·7}{1·08}=181 < 350$

f_c=335·5 kg/cm²

Increased f_c=1·33×335·5=446 kg/cm²

Safe load=446×3·11=1390 kg > 1025 kg

3·7·5 Member 11

Member is in compression
Maximum compression=1825 kg
Maximum tension=·85 kg
Length of member=251·5 cm

Try ID—32 mm—M,

A=4·0 cm², r=1·39 cm

$\frac{l_e}{r}=\frac{251·5\times0·7}{1·39}=127$

l/r=120, f_c=674 kg/cm²
l/r=130, f_c=603 kg/cm²
l/r=127, f_c=625 kg/cm²

Safe load$=625\times4=2500$ kg $>$ 1825 kg

Area required for tension$=\frac{785}{1250}=0{\cdot}625$ cm^2

$<$ 4 cm^2 provided.

Use 32 mm *ID*—Medium section.

3·8 Column

Maximum load on column due to DL+LL

$=1491+2110=3601$ kg

Self wt. @ 10 kg/m$=42{\cdot}5$

Total load$=3643{\cdot}5$

Say$=3{\cdot}7$ tonnes

Length of column $=425$ cm

Least radius of gyration required $(l/r < 180)$

$$r=\frac{425\times0{\cdot}85}{180}=2{\cdot}0 \text{ cm}$$

Try 50 mm L_2

Area$=5{\cdot}82$ cm^2, $r=2{\cdot}02$ cm.

$$\frac{l_e}{r}=\frac{425\times0{\cdot}85}{2{\cdot}02}=178$$

For $l/r=178$, $f_c=347{\cdot}4$ kg/cm^2

Safe load$=347{\cdot}4\times5{\cdot}82=2030$ kg$=2{\cdot}03$ tonnes $<$ **3·7 tonnes**

Hence unsafe.

Try 80—L

$A=8{\cdot}74$ cm^2, $r=3{\cdot}03$ cm $>$ 2·0 cm required

$$\frac{l_e}{r}=\frac{42\times0{\cdot}855}{3{\cdot}03}=119<180$$

for $l/r=119$, $f_c=681{\cdot}1$ kg/cm^2

Safe load$=681{\cdot}1\times8{\cdot}74=5990$ kg$=5{\cdot}99$ tonnes $>$ **3·7 tonnes**

uneconomical.

Try 65 mm L

$A=7{\cdot}44$ cm^2, $r=2{\cdot}58$ cm

$$\frac{l_e}{r}=\frac{425\times0{\cdot}85}{2{\cdot}58}=144<180$$

for $l/r=144$. $f_c=510$ kg/cm^2

Safe load$=510\times7{\cdot}44=3800$ kg$=3{\cdot}8$ tonnes $>$ 3·7 tonnes.

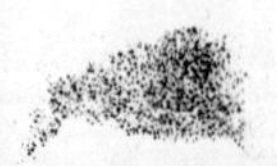

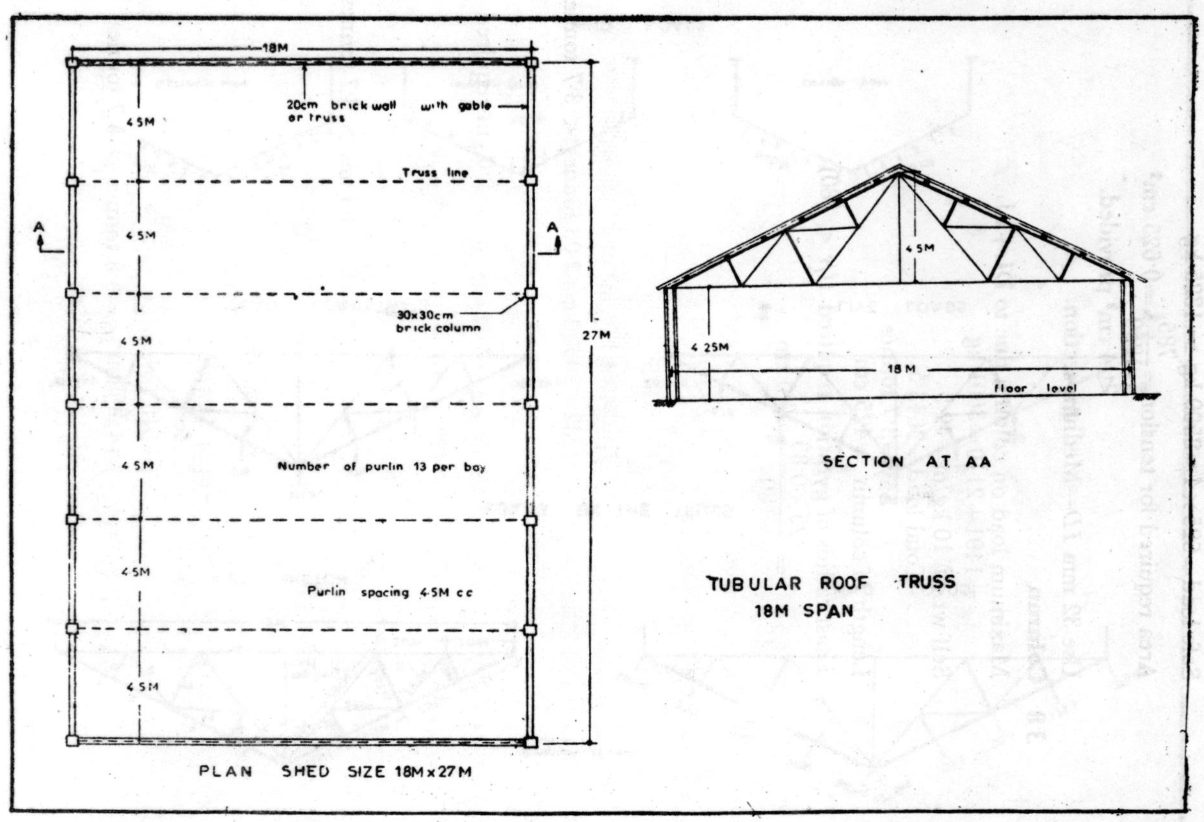
18M
20cm brick wall with gable or truss
Truss line
A
A
30x30cm brick column
27M
4·5M
4·5M
4·5M
4·5M
4·5M
4·5M
Number of purlin 13 per bay
Purlin spacing 4·5M cc
PLAN SHED SIZE 18M x 27M
4·5M
4·25M
18 M
floor level
SECTION AT AA
TUBULAR ROOF TRUSS
18M SPAN

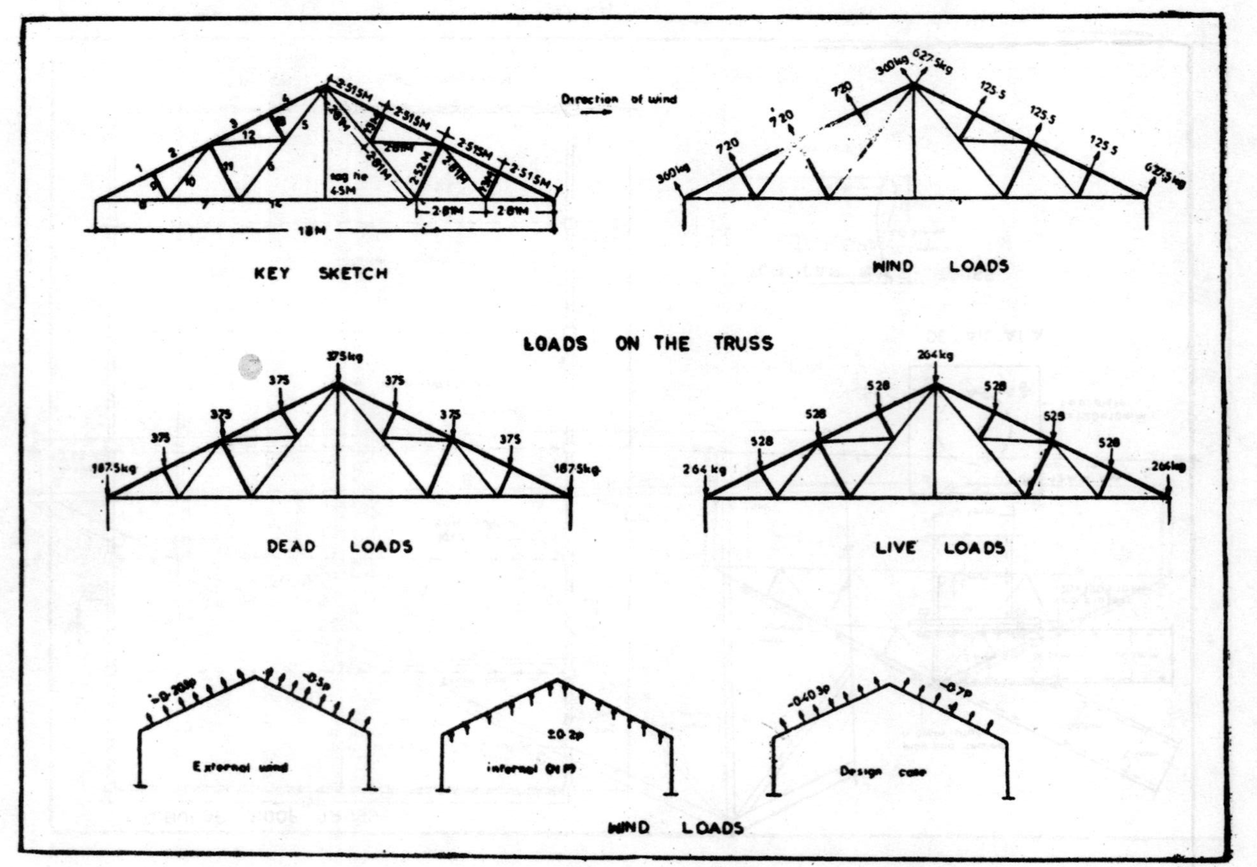

LOADS ON THE TRUSS
KEY SKETCH
18M
2·81M
2·515M
Direction of wind
WIND LOADS
360kg
627·5kg
720
125·5
DEAD LOADS
375kg
375
187·5kg
LIVE LOADS
264kg
528
External wind
-0·5p
Design case
-0·7p
WIND LOADS

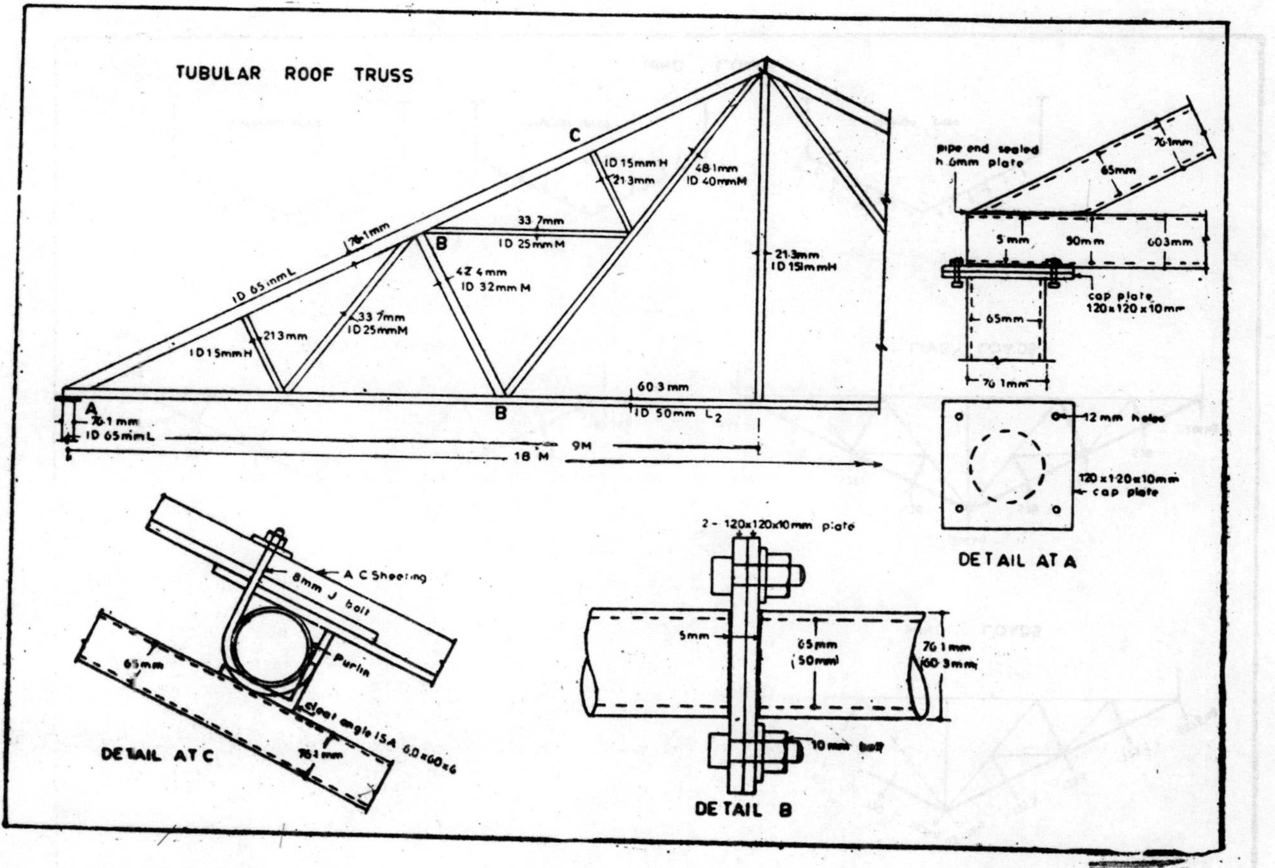
TUBULAR ROOF TRUSS
C
ID 15mm H
21.3mm
48.1mm
ID 40mmM
33.7mm
ID 25mmM
B
76.1mm
ID 65mmL
42.4mm
ID 32mmM
33.7mm
ID 25mmM
21.3mm
ID 15mmH
21.3mm
ID 15mmH
60.3mm
ID 50mm L2
A
76.1mm
ID 65mmL
B
9M
18 M
pipe end sealed
h 6mm plate
76.1mm
65mm
5mm
50mm
60.3mm
cap plate
120x120x10mm
65mm
76.1mm
12mm holes
120x120x10mm
cap plate
DETAIL AT A
A C Sheeting
8mm J bolt
Purlin
65mm
cleat angle ISA 60x60x6
76.1mm
DETAIL AT C
2 - 120x120x10mm plate
5mm
65mm
(50mm)
76.1mm
(60.3mm)
10mm bolt
DETAIL B

IX. Design of Riveted Plate Girders

1

Riveted Plate Girders —General Information

roduction

Plate girders are used when standard rolled sections do not provide the required section modulus. Building up a beam by riveting plates and angles to increase the modulus of section is one of the choices available to the designer. Plate girders are especially adopted on short spans and heavy loads. These girders can be used for buildings, highway and railway bridges.

1·2 Components

In a riveted plate girder the components are :

(*i*) Web plate (*ii*) Flange plates (*iii*) Flange angles

(*iv*) Stiffners (*a*) Intermediate stiffners (*b*) Bearing stiffners
(*c*) Longitudinal stiffners

(*v*) Splicing

The two main components of the plate girders namely flanges and webs are designed to resist bending moment and shear force respectively.

1·3 Effective Span

The effective span for plate girder is usually the centre to centre of end bearings.

Clear span Metres	Effective span Metres
3	3·3
6	6·45
9	9·52
12	12·60
15	15·675
18	18.75
21	21·825
24	24·90

1·4 Proportioning of Components

Depth of Girder

The approximate depth of the girder is usually taken as $\frac{1}{8}$ to $\frac{1}{12}$ of the effective span.

For lighter loads—$\frac{L}{12}$

Medium loads—$\frac{L}{10}$

Heavy loads—$\frac{L}{8}$

The economical depth of beam $d=5\cdot5\sqrt[3]{\frac{M}{f_b}}$

1·4·1 Width of Flange

Width of flange should be within the range of $\frac{L}{40}$ to $\frac{L}{45}$.

The maximum width given by this rule may be exceeded provided the flange plates are adequately stiffened.

1·5 Self Weight

The self weight can be estimated by computing weight of flange plates with angles and web plates and adding 50% of web weight for stiffners etc.

1·6 Limiting Deflection

The maximum deflection should not exceed $\frac{1}{325}$ of the span.

1·7 Allowable Stresses

Upto 20 mm web thickness

bending stress=1575 kg/cm²

shear stress =945 kg/cm²

over 20 mm thick web

bending stress=1500 kg/cm²

shear stress =865 kg/cm²

1·8 Minimum Thickness of Plates

For those girders exposed to weather but accessible for painting =6 mm

For those girders exposed to whether but not accessible for painting =8 mm

For highway bridges=8 mm

For railway bridges =10 mm

Analysis

Moment of resistance of plate girders can be obtained by two methods.

1. Flange area or chord stress method
2. The moment of Inertia method

For analysing a plate girder section the moment of Inertia of the gross-section may be used to find the stresses and then the stresses may be increased on the tension flange in the ratio of its gross area to its net area.

Plate girder is a built up beam and moment of resistance can be found out by flexure formula :

$M=f\,I/y$

M=Moment of resistance

$I=M.I.$ of girder

f=Max. allowable stress

y=extreme fibre distance.

Flange area method

This method is approximate but is sufficiently accurate for practical purposes.

F=Resultant compressive and tensile force

D=Distance between lines of action of these forces.

Moment of girder section$=F\times D$

$M=FD$

if f_b=average bending stress

A=flange area

then $F=f\times A$

Hence $M=f\times A\times D$

But this does not take the Bending Moment contributed by the web and the error is on the safe side.

However if the value for f is the maximum allowable fibre stress, the error is on the unsafe side, since f in the above formula is the resultant stress at the C.G. of the flange.

D=the distance between the centres of gravity of tension and compression flanges, and is taken as the web depth or slightly less.

Since the flanges of the plate girder are narrow as compared to the total depth of girder the fibre stress may be assumed uniform across the flange section.

Let f_b=average uniform stress on flange area.

=fibre stress at centre of gravity of flange area.

A_g=Gross area of one flange (angles+cover plates)

D=distance between C.G. of flanges=effective girder depth

A_w=Area of web plates$=h\times t_w$

M_f=resisting moment of the flange

M_w=resisting moment of the web plate

$M_f=A_g\times f_b\times D$

Assuming that the fibre stress at the edges of the web plate is f_b we have M_w=Area×Average stress×Lever arm

$$M_w=t_w\times h/2\times f_b/2\times 2h/3$$
$$=1/6\, f_b h^2 t_w$$

Total resisting moment $M=M_f+M_w$

$$=A_g\times f_b\times D+1/6\, f_b h^2 t_w$$

Since D is usually taken as h, and $h\times t_w=A_w$

We can write

$$M=A_g f_b D+1/6\, f_b h t_w\times D=A_g f_b D+1/6\, f_b A_w\times D$$
$$=f_b D(A_g+1/6\, A_w)=f_b AD$$

Where $A=A_g+1/6\, A_w$

There will be no appreciable error if the depth of web plate and effective depth are considered equal which is practically the case in most girders.

$$M=f_b\times D(A_g+1/6\, A_w)$$

$1/6\, A_w$ is called the web equivalent

For tension flange

Net area is to be taken for calculations. Area of the rivet holes must be deducted. Usually 22 mm ϕ rivets at 10 mm is comm...

Net effective web equivalent$=\frac{3}{4}\times\frac{1}{6}A_w=\frac{1}{8}A_w$

If A_n=net area of the flange, then the resisting moment of girder $M=f_b D(A_n+\frac{1}{8}A_w)$.

These formulae are on slightly unsafe side if f_b is regarded as the maximum allowable fibre stress.

1·9 Specifications For Web Plate

Minimum thickness of web plate for unstiffened webs$=\frac{d}{85}$

For vertically stiffened webs$=\frac{1}{180}$ of the smallest clear panel dimension$=\frac{1}{200}$ for webs with longitudinal stiffeners

Stiffeners at $\frac{2}{5}$ d from the compression flange$=\frac{d}{250}$

For webs with longitudinal stiffners at N.A.$=\frac{d}{400}$

1·9·1 Flange Plates

Maximum outstand for compression flange $=16\ t$

Maximum outstand for tension flange $=20\ t$

Flange angles shall be at least $\frac{1}{3}$ of the gross flange area.

1·10 Curtailment of Flange Plates

Since the maximum bending moment occurs only at centre, the designed cross-section is only required at centre. As the B.M. reduces from centre towards supports the section designed for maximum. B.M. can be reduced. This reduction can be met with cutting out top plates at some points on the girder. This method of proportioning flange plates, to achieve economy in the design, is called the curtailment of flange plates.

1. Graphical method
2. Analytical method

(1) Superimposition of moment of resistance diagram over B.M. diagram.

1·10·1 Analytical Method

Let the girder with span L carry a udl w

M=B.M. @ centre of the span

M_x=B.M. @ any point K at a distance x from centre

A=flange area required at the centre

A_x=flange area required at point K

w=load per unit length of span

L=effective span

Since entire B.M. is taken up by flanges we can say that flange area is proportional to B.M.

$$\frac{A}{A_x}=\frac{M}{M_x}$$

$$\frac{A-A_x}{A}=\frac{M-M_x}{M} \qquad \ldots($$

$$M=\frac{wL^2}{8}$$

$$M_x=\frac{wL}{2}\left(\frac{L}{2}-x\right)-\frac{w}{2}\left(\frac{L}{2}-x\right)$$

$$M_x=\frac{WL^2}{8}-\frac{wx^2}{2}$$

Substituing $\frac{wL^2}{8}=M$

$$M-M_x=\frac{wx^2}{2}$$

$$\frac{M-M_x}{M}=\frac{wx^2}{2}\Big/\frac{wL^2}{8}=\frac{4x^2}{L^2} \qquad \ldots(2)$$

From (1) and (2)

$$\therefore \quad \frac{4x^2}{L^2} = \frac{A-A_x}{A}$$

$$2x = L\sqrt{\frac{A-A_x}{A}}$$

Let $2x = l$ = length of plate or distance at which the flange plate can be curtailed

$$l = L\sqrt{\frac{A-A_x}{A}}$$

$A-A_x$=area of the cover plate or flange plate which is to be curtailed

$$A-A_x = a, \text{ then } l = L\sqrt{\frac{a}{A}}$$

or

$$l_1 = L\sqrt{\frac{a_1}{A}}$$

In general if 3 plates of areas a_1, a_2, a_3 are used then length of plates required l_1, l_2, l_3 are found similarly.

$$\text{as } l_1 = L\sqrt{\frac{a_1}{A}}$$

$$l_2 = L\sqrt{\frac{a_1+a_2}{A}}$$

$$l_3 = L\sqrt{\frac{a_1+a_2+a_3}{A}}$$

or

$$l_n = L\sqrt{\frac{a_n}{A}}$$

Specifications also state that first cover plate should not be curtailed.

Each flange plate should be extended beyond its theoretical cut off point and the extension shall contain sufficient rivets to develop in the plate, the load calculated for the B.M. in the girder at the theoretical cut-off point.

1·11 Stiffners

Intermediate stiffners.

Intermediate stiffners are required to prevent the web plate from buckling under a complex and variable stress distribution resulting from combined shear and bending moment.

The stiffners have a secondary function not generally recognised in that if fitted against the flanges at top and bottom, they maintain the perfect 90° angle between flanges and the web.

These stiffners are provided when d/t_w 85

Spacing of the stiffness $C \nless d/3$ and $\ngtr 1{\cdot}5d$

Moment of inertia of the stiffner $I_s \nless \frac{1 \cdot 5\, d^3\, t^3}{C^2}$

The width of the stiffner should not exceed 16 times the thickness of the stiffner.

d=distance between the flange angles.

Intermediate vertical stiffners may be in pairs arranged one on each side of the web or may be single. When single stiffers are used, they should preferably be placed alternately on opposite sides of the web. The stiffners shall extend from flange to flange and may be joggled over the vertical legs of the flange angles.

1·12 Web Panel Dimensions

In no case shall the greater unsupported clear dimensions of a web panel exceed 270 t_w nor the lesser unsupported clear dimensions of the same panel exceed 180 t_{ws} where t_w=thickness of web plate.

1·13 Outstand of the Stiffners

For sections other than flats, the outstand of all the stiffners from the web shall not be more than 16t.

For flats outstand shall not be greater than 12 t.

Where t is the thickness of flat or section.

1·14 Bearing Stiffners

The function of the bearing stiffners is to transmit concentration of load so as to avoid local bending failure of the flanges and local crippling or buckling of the web.

Selection of stiffners is usually made on the basis of local permissible contact bearing pressure of 1890 kg/cm² at the points of the bearing stiffners and the flanges.

Rivets shall be used to transfer the total load from the bearing stiffners to the web.

The bearing stiffners together with the web plate shall be designed as a column with an equivalent reduced slenderness ratio.

Effective length=0·7×Actual length.

The area of the section which resists compression is the pair of stiffners together with a length of web on each side of the centre line of the stiffners when available=20 times the web thickness.

The bearing stiffners at the support points should project as nearly as possible to the outer edges of the flanges and filler plates shall be used.

1·15 Longitudinal Stiffners

In addition to the vertical intermediate stiffners, longitudinal stiffners shall be provided if $t_w < \frac{d}{200}$.

One horizontal stiffner shall be placed on the web at a distance from the compression flange=2/5 of the distance from the compression flange to the neutral axis. When the thickness of the web is $< \frac{t_w}{250}$. These stiffners shall be designed so that $I \nless 4C_1\, t^3$. When $d/t_w > 250$ a second stiffner is placed at N.A. for which $I \nless dt^3$ where t is the thickness of the stiffner.

Horizontal web stiffners shall extend between vertical stiffners but need not be continuous over them.

Horizontal stiffners may be in pairs arranged on either side of the web or may be single.

1·16 Connections (Flange Plate to Flange Angles)

$$\text{Pitch of rivets} = \frac{n \times \text{Governing strength of rivet} \times I}{F\, A\bar{y}}.$$

where n=No. of rivets in one pitch length

I=moment of inertia

$A\bar{y}$=Moment of the net area of flange plate about NA

F=shear force at the section.

1·16·1 Web Plate and Flange Angle

$$\text{Pitch of rivet } p = \frac{n \times \text{Governing strength of rivet} \times I}{F\, A\bar{y}}$$

$A\bar{y}$=Moment of the net area of the flange angle and flange plates about NA.

1·16·2 Intermediate Stiffners to Web

The stiffner connection to web plate shall be designed to develop a shearing force in tonnes per cm run of not less than $\frac{t^2_w}{2h}$.

where t_w=Thickness of web plate in cm,

h=Outstand leg width of stiffner.

1·16·3 Bearing Stiffner to Web

Connection shall be designed to transmit the entire concentrated load to the web.

1·17 Splices

(1) Web splice, (2) Flange plate splice, (3) Flange angle splice.

1·17·1 Web Splice

Since there is a limit to the length and weight of plate, the webs of long span girders must consist of two or more pieces.

For riveted girders, the splice consists of two plates one in each side of the web.

The web splice can best be located at a section, where there is enough excess flange plates, if this cannot be done, then splice plates be added to the flanges.

Some engineers argue it is unwise to splice a web plate at the section of maximum moment, but it would be seem that an inadequate splice at a section of smaller moment could be just as fatal as could one at the section of maximum moment.

Web splice should be designed for the full strength of web in both shear and bending.

2nd Type

The plates 'A' called moment plates to resist the web's share of the bending moment and the plates 'B' called shear plates to resist the shear are used in this type.

Size of the moment plates, is decided on the No. of rivets to be accommodated.

Size of the shear plates can be determined on the basis of shear area required.

Design of plates "A"

Let f=allowable stress in bending in each flange

Let h=Distance between C.G. of the flanges

A_w=Area of the web

h_1=Distance between C.G. of outer plates A

F=Total force on each pair of the plates A

S=Maximum shear at the splice

Moment resisted by web$=\frac{1}{8}A_w f\times h$

Moment of resistance of plates$=Fh_1$

Since the plates A, are to be designed to take up the moment of resistance of the web.

$$Fh_1=\frac{1}{8}A_w fh$$

$$F=\frac{1}{8}\,\frac{A_w\times fh}{h_1}$$

Therefore the plates are to be designed for the above value.

Let d_1=Depth of the plates A

t_1=Thickness of the plates A

Required area for each pair of plates $A=\dfrac{F}{f_1}$

Where f_1=Allowable stress in bending for the plates,

f_1 may be taken as$=\dfrac{h_1}{h}\times f$

But the area of one pair of plates $=2\times d_1\times t_1$

Assuming suitable thickness t_1, $d_1=\dfrac{A}{2t_1}$

t_1 is limited to the thickness of web.

Total No. of rivets required on each side of the splice, to connect each pair of plates *A*

$$\frac{F}{r_1} \quad \text{where } r_1=\frac{h_1}{h}\,r$$

where r=least rivet value.

Width

Width of the plate is selected with respect to No. of rivets to be accommodated, a minimum pitch of 2·5 d should be used.

A maximum pitch $\not>$ 32 t or 300 mm

1·18 Design of Central Plate

Central plates 'B' consist of two plates one on each side of the web. These plates are to be designed to resist the maximum shear force.

The total area of the plates $B=\frac{S}{f_s}$

where f_s=allowable stress in shear

By assuming a suitable thickness t_2, d_2 can be found as $d_2=\frac{S}{2_s f t_2}$ and the thickness is limited to the web thickness.

2

A Riveted Plate Girder For a Cinema Balcony

2·1 Data

Span =20 m
Loading =Balcony Floor
Cross girder spacing =4 m c/c
RCC Slab thickness =15 cm
Span for cross girder =10 m
Self weight of cross girder=200 kg/m

Interior cross girders transmit load coming on the area of 20 sqm as concentrated load on the plate girder. No horizontal stiffners to be provided.

2·2 Relevant Codes

IS—800

2·3 Loads

Live load=500 kg/m²
Self weight of RC slab (15 cm)=15×24=360 kg/m²
Dead load due to seating and steps etc=140 kg/m²
Total load=1000 kg/m²
Load transferred by the cross girder as reaction on plate girder
=1000×20=20,000 kg=20 tonnes
Self wt. of cross girder=200×5=1000 kg=1 tonne.
Total reaction from interior cross girder to plate girder
=21 tonnes.
Total reaction from end cross girder

$$=\frac{20}{2}+1=10+1=11 \text{ tonnes}$$

2·3·1 Self Weight of Plate Girder

$$\text{Self weight } \frac{W}{300}=\frac{\text{Total load on the girder}}{300}$$

$$=\frac{21\times4+11\times2}{300}=\frac{106}{300}$$

$$=0{\cdot}353 \text{ tonnes/m Say } 350 \text{ kg/m.}$$

2·3·2 Loading on Plate Girders

The girder carries 4 concentrated loads from the cross girders and uniformly distributed load of 0·35 tonnes/m due to its own weight.

2·3·3 Bending Moment

$$\text{Reaction}=\frac{84}{2}+\frac{20\times0{\cdot}35}{2}=42+3{\cdot}5=45{\cdot}5 \text{ tonnes}$$

Bending moment at centre

$$M_c=45{\cdot}5\times10-21\times6-21\times2-0{\cdot}35\times10\times5$$
$$=455-126-42-17{\cdot}5=455-185{\cdot}5=269{\cdot}5 \text{ m tonnes}$$

2·3·4 Shear Force

$$\text{Shear Force due to DL+LL}=\frac{21\times4+11\times2}{2}=53 \text{ tonnes}$$

$$\text{Due to self weight}=\frac{0{\cdot}35\times20}{2}=3{\cdot}5 \text{ tonnes}$$

Total SF=S=56·5 tonnes

2·3·5 Web Plates

$$\text{Approximate depth of girder}=\frac{1}{12} \text{ of span}$$
$$=\frac{1}{12}\times2000=166 \text{ cm}$$

$$\text{Economical depth}=d_w=5{\cdot}5\sqrt[3]{\frac{M}{f}}$$
$$=5{\cdot}5\sqrt[3]{\frac{269{\cdot}5\times100\times1000}{1500}}=144 \text{ cm}$$

Try d_w=140 cm.

2·3·6 Web Thickness

Since horizontal stiffners are not to be provided

$$t_w \nless \frac{d_c}{200}$$

where d_c=clear distance between the flange angles

Using ISA 150×150×10 mm

$$d_c=(140+0{\cdot}5\times2)-2\times15=141-30=111 \text{ cm}$$
$$t_w=\frac{111}{200}=0{\cdot}555 \text{ cm}=5{\cdot}5 \text{ mm}$$

Use minimum thickness of web plate=6 mm

2·3·7 Web Thickness on Shear Consideration

$$t_w=\frac{S}{q_s\times d_w}=\frac{56500}{945\times140}={\cdot}42 \text{ cm}$$

Use 6 mm web plate

2·3·8 Flange Plates

$$\text{Width of flange}=\frac{\text{span}}{40} \text{ to } \frac{\text{span}}{50}=\frac{200}{40} \text{ to } \frac{200}{50}$$
$$=50 \text{ cm to } 40 \text{ cm}$$

Try flange width=40 cm

Approximate area of flange required

$$M_c = F_b \times A \times D$$

$$A = \frac{M_c}{f_b D} = \frac{269{\cdot}5 \times 100 \times 1000}{1500 \times 141} = 127 \text{ cm}^2$$

2·4 Trial Section

Angles 2 Nos. ISA $150 \times 150 \times 10 = 2 \times 29{\cdot}03 = 58{\cdot}06$ cm²
plates 2 Nos. 10 mm thick $= 2 \times 1 \times 40 = 80{\cdot}00$ cm²
$\frac{1}{8}$ web area $= \frac{1}{8} \times 0{\cdot}6 \times 11' = 8{\cdot}35$ cm²
Total area $= 146{\cdot}41$ cm²

Deduct for rivet holes in tension flange

Using 22 mm ϕ rivets for connection

Angles—2 holes $= 2 \times 2{\cdot}35 \times 1 = 4{\cdot}70$ cm²
Flanges—2 holes (Zig-Zag riveting) $= 2 \times 2{\cdot}35 \times 3$
$= 14{\cdot}10$ cm²
Total $= 18{\cdot}80$ cm²

Net flange area $= 146{\cdot}41 - 18{\cdot}80 = 128{\cdot}61$ cm² > 127 cm² required.

2·4·1 Stresses

I_{xx} of the web plate $= \frac{t_w \times d^3{}_w}{12} = \frac{0{\cdot}6 \times 140^2}{12} = 148{,}000$ cm⁴

I_{xx} for flange angles $= 2 \times I_{xx}$ of angles about their own axis $+$ Area $\times h^2{}_f$

$$= 2\left\{\left(2 \times 622{\cdot}4 + 29{\cdot}03\right) \times \left(\frac{141}{2} - 4{\cdot}06\right)^2\right\}$$

$$= 2\ (1244\ 8 + 29{\cdot}03 \times 66{\cdot}44^2) = 518,\ 450 \text{ cm}^4$$

I_{xx} for plates $= I_{xx}$ of plates about their own axis $+ Ah^2{}_p$

$$= 2\left[\frac{1}{12} \times 40 \times 2^3 + 40 \times 2 \times \left(\frac{145}{2}\right)^2\right] = 2(26{\cdot}4 + 80 \times 72{\cdot}5^2)$$

$$= 2(26{\cdot}4 + 406000) = 812052{\cdot}8 \text{ cm}^4$$

Total $I_{xx} = 148{,}000 + 518450 + 812053 = 1478503$ cm⁴
$= 1{\cdot}4785 \times 10^6$ cm⁴

Deduct for rivet holes in :
flange plates $= 2(2 \times 2{\cdot}35 \times 3 \times 71^2) = 143{,}000$ cm⁴
flange angles $= 2(2{\cdot}35 \times 2{\cdot}36 \times 62{\cdot}5^2) = 47600$ cm⁴
Total I_{xx} for holes $= 190600$ cm⁴
Net I_{xx} for the trial section $= 1{\cdot}4785 \times 10^6 - {\cdot}19 \times 10^6$
$= 1{\cdot}2885 \times 10^6$ cm⁴ $= 1{\cdot}29 \times 10^6$ cm⁴

Maximum fibre stress in the tension flange

$$f_h = \frac{M_c}{Z} = \frac{269{\cdot}5 \times 100 \times 1000 \times 72{\cdot}5}{1\ 29 \times 10^6}$$

$$= 1520 \text{ kg/cm}^2 < 1575 \text{ kg/cm}^2$$

The assumed trial section is safe.

2·5 Curtailment of Flange Plates

$$l_1 = L\sqrt{\frac{a_1}{A}} = 20\sqrt{\frac{40 \times 1}{128{\cdot}61}} = 10{\cdot}7 \text{ m}$$

Top flange plate can be curtailed at a distance

$$=\frac{10\cdot7}{2}+\cdot3=5\cdot35+0\cdot3=5\cdot65 \text{ m from centre of the beam}$$

2·6 Connection of Flange Plate to Angles

Maximum shear forces $=S=56500$ kg

vertical shear/cm height $=\frac{S}{h}=\frac{56500}{(141-30)}=508$ kg

Least strength of 22 ϕ rivet in single shear

$$=\pi\times\frac{2\cdot35^2\times1025}{4}=4446 \text{ kg}$$

If Zig-Zag riveting is used at support the rivets are in single shear as there is only one plate, since the top plate is curtailed Strength of rivet in bearing (10 mm thick) $=2\cdot35\times1\times2360=7764$ kg

Pitch of rivets $=\frac{4446}{508}=8\cdot8$ cm

Maximum pitch allowed $=12\times t=12\times1=12$ cm c/c

Minimum pitch $=2\cdot5\times d=2\cdot5\times2\cdot2=5\cdot5$ cm c/c

Use 22 mm ϕ at 7·5 c/c Zig-Zag

2·6·1 Connection of Flange

The rivets are in double shear

Least strength of rivet in Double Shear $=4446\times2=8892$ kg

Strength in bearing in web plate $=0\cdot6\times2\cdot35\times2360=3340$ kg

Pitch of the rivets $=\frac{3340}{508}=6\cdot6$ cm

Use 22 mm ϕ at 6 mm c/c

2·7 Stiffners

$$\frac{d_w}{t_w}=\frac{111}{0\cdot6}=186>85$$

Intermediate stiffners shall be provided

Spacing of the stiffners $=C=\frac{d_w}{3}$ to $1\cdot5\ d_w$

$$=\frac{111}{3} \text{ to } 1\cdot5\times111=37 \text{ to } 165 \text{ cm}$$

2·7·1 Panel Dimensions

Larger dimension of the panel

$270\times t_w=270\times0\cdot6=182$ cm

Shorter dimension of the panel $=180\times t_w=180\times0\cdot6=108$ cm

Use stiffner spacing $C=100$ cm

2·7·2 Design of Intermediate Stiffner

Try ISA $100\times75\times10$ mm with packing plate of 10 mm thick.

Moment of Inertia of the stiffners required $=\frac{1\cdot5\times d^3_w\times t^3_w}{C^2}$

$$=\frac{1\cdot5\times111^3\times1^3}{100\times100}=205 \text{ cm}^4$$

I_{xx} of the stiffners about centre line of web

$=16\cdot5\ (3\cdot19+1+0\cdot3)^2+160\cdot4=455\cdot4\ \text{cm}^4>205\ \text{cm}^4$

Use ISI 100×75×10 mm at cm c/c as intermediate stiffner.

2·7·3 Connection

The rivets connecting intermediate stiffner shall transmit a shear force equal to $\frac{t^2}{2h}$ tonnes/cm length of stiffner.

$$\text{Shear/cm length}=\frac{0\cdot6^2}{2\times(10-1)}=\frac{0\cdot36\times1000}{18}=20\ \text{kg/cm}$$

Shear to be transmitted is negligible, therefore use 18 mm ϕ rivets at 10 cm c/c.

2·7·4 Bearing Stiffner—End Stiffner

Maximum shear force at support=56500 kg

$$\text{Bearing area required}=\frac{56500}{1890}=30\ \text{cm}^2$$

Try ISA 100×75×10 mm 4 Nos.

Area provided $=4(10-1\cdot1)\times1=4\times8\cdot9\times1=35\cdot6\ \text{cm}^2>29\cdot5\ \text{cm}^2$

Load on the column=56500 kg

Length of column =111 cm

Effective length =0·7×111=77·7 cm

$I_{xx}=4[16\cdot5\times(3\cdot19+1\cdot3)^2+160\cdot4]=4(16\cdot5\times4\cdot22^3+160\cdot4)$

$=4\times455=1820\ \text{cm}^4$

$\text{Area}=16\cdot5\times4+4\times1\times7\cdot5+40\times0\cdot6+20\times0\cdot6\times0\cdot6=127\cdot2\ \text{cm}^2$

$$r=\sqrt{I/A}=\sqrt{\frac{1820}{127\cdot2}}=3\cdot8\ \text{cm}$$

$$l/r=\frac{77\cdot7}{3\cdot8}=20\cdot4$$

Allowable stress=1225 kg/cm²

Safe load=1225×127 2=157000 kg=157 tonnes>56·5 tonnes.

2.7·5 Under Concentrated Loads

Load on the stiffner=21 tonnes

$$\text{Bearing area required}=\frac{21000}{1890}=11\cdot1\ \text{cm}^2$$

Try ISA 100×75×10 mm 2 Nos, one on either side of the web

Area provided $=2(10-1\cdot1)\times1=2\times8\cdot9\times1=17\cdot8\ \text{cm}^2$

$I_{xx}=2[160\cdot4+16\cdot5\ (3\cdot19+1+0\cdot3)^2]=2\times455=910\ \text{cm}^4$

$\text{Area}=2\times16\cdot5+2\times20t_w\times t_w=33+2\times20\times0\cdot6\times0\cdot6$

$=33+14\cdot4=57\cdot4\ \text{cm}^2$

$$r=\sqrt{\frac{I}{A}}=\sqrt{\frac{910}{57\cdot4}}=\sqrt{15\cdot8}=3\cdot95\ \text{cm}$$

$$l/r=\frac{111\times0\cdot7}{3\cdot95}=19\cdot7$$

Allowable stress$=1225$ kg/cm^2

Safe load$=1{\cdot}225\times57{\cdot}4=71$ tonnes>21 tonnes.

2·8 Web Splice

Usually plates are available in maximum lengths of 12 m. Since the span of the girder is 20 m, two lengths are to be used to make up this length.

Splice is located at 3 m from one end. The splice has to resist shear at that section and also bending moment carried by the web at that section.

2·8·1 Moment Plates

Moment to be resisted by the web$=\dfrac{A_w\, f_b\times d}{8}$

$$=\frac{1}{8}\times0{\cdot}6\times111\times1575\times145$$

$M_w=1{\cdot}9\times10^6$ cm kg

Assuming two rows of 22 mm ϕ rivets in the moment plates, width of plate$=2{\cdot}5d\times2+2\times$edge distance

$$=2{\cdot}5\times2{\cdot}35\times2+2\times3{\cdot}8=11{\cdot}7+7{\cdot}6=19{\cdot}3 \text{ cm}$$

Try 2—200 mm wide plate

$d_1=111-20=91$ cm

Force to resisted by the moment plate

$$F=\frac{M_w}{d}=\frac{1{\cdot}9\times10^6}{91}=20{\cdot}9\times10^3=20{,}900 \text{ kg}$$

$$\text{Area required}=\frac{\text{Force}}{\text{allowable stress}}=\frac{F}{f_b'}$$

$$f_b'=f_b\times\frac{d_1}{d}=1575\times\frac{91}{145}=990 \text{ kg/cm}^2$$

$$\text{Area of moment plate}=\frac{20900}{990}=21 \text{ cm}^2$$

Area for each piate$=10{\cdot}5$ cm^2

Area of plate provided$=0{\cdot}6\times20=12$ cm$^2>10{\cdot}5$ cm^2

$$\text{No. of rivets required}=\frac{\text{Force}}{\text{Least strength of rivet}}$$

Strength of rivet in single shear (22 mm ϕ)$=4446$ kg

Strength in bearing. (0 6 mm thick)$=3310$ kg

Least strength of rivet at the C.G. level of moment plate

$$=3340\times\frac{91}{145}=2100 \text{ kg}$$

$$\text{Number of rivets required}=\frac{20900}{2100}\simeq10$$

Provide 20 mm ϕ in two row s of 5 each on either side of the splice.

2·8·2 Shear Plate

Shear force at the section

$S_s = 45.5 - 21 - 0{\cdot}35 \times 8 = 45{\cdot}5 - 21 - 2{\cdot}8 = 21{\cdot}7$ tonnes

Area of the shear plate required $= \dfrac{S_s}{f_s} = \dfrac{21700}{945} = 23$ cm^2

Depth of plate available $= 111 - 40 = 71$ cm

Use 70 cm depth.

Thickness of the plate $= 6$ mm

Area provided $= 70 \times {\cdot}6 = 42$ cm$^2 > 23$ cm^2

Number of rivets required $= \dfrac{21700}{3340} \simeq 7$

Use 8 rivets @ 8·5 cm c/c in vertical row on either side of the splice.

TABLES

TABLE I ROLLED STEEL BEAMS

DIMENSIONS AND PROPERTIES

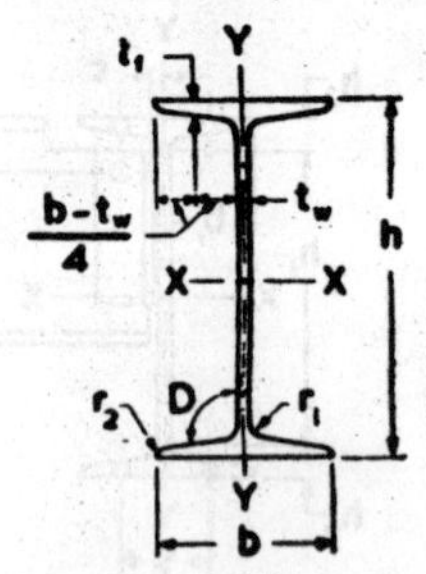

Designation	Weight per Metre	Sectional Area	Depth of Section	Width of Flange	Thick-ness of Flange	Thick-ness of Web	Moments of Inertia		Radii of Gyration	
	w	a	h	b	t_f	t_w	I_{xx}	I_{yy}	r_{xx}	r_{yy}
	kg	cm^2	mm	mm	mm	mm	cm^4	cm^4	cm	cm
ISJB 150	7.1	9.01	150	50	4.6	3.0	322.1	9.2	5.98	1.01
ISJB 175	8.1	10.28	175	50	4.8	3.2	479.3	9.7	6.83	0.97
ISJB 200	9.9	12.64	200	60	5.0	3.4	780.7	17.3	7.86	1.17
ISJB 225	12.8	16.28	225	80	5.0	3.7	1 308.5	40.5	8.97	1.58
ISLB 75	6.1	7.71	75	50	5.0	3.7	72.7	10.0	3.07	1.14
ISLB 100	8.0	10.21	100	50	6.4	4.0	168.0	12.7	4.06	1.12
ISLB 125	11.9	15.12	125	75	6.5	4.4	406.8	43.4	5.19	1.69
ISLB 150	14.2	18.08	150	80	6.8	4.8	688.2	55.2	6.17	1.75
ISLB 175	16.7	21.30	175	90	6.9	5.1	1 096.2	79.6	7.17	1.93
ISLB 200	19.8	25.27	200	100	7.3	5.4	1 696.6	115.4	8.19	2.13
ISLB 225	23.5	29.92	225	100	8.6	5.8	2 501.9	112.7	9.15	1.94
ISLB 250	27.9	35.53	250	125	8.2	6.1	3 717.8	193.4	10.23	2.33
ISLB 275	33.0	42.02	275	140	8.8	6.4	5 375.3	287.0	11.31	2.61
ISLB 300	37.7	48.08	300	150	9.4	6.7	7 332.9	376.2	12.35	2.80
ISLB 325	43.1	54.90	325	165	9.8	7.0	9 874.6	510.8	13.41	3.05
ISLB 350	49.5	63.01	350	165	11.4	7.4	13 158.3	631.9	14.45	3.17
ISLB 400	56.9	72.43	400	165	12.5	8.0	19 306.3	716.4	16.33	3.15
ISLB 450	65.3	83.14	450	170	13.4	8.6	27 536.1	853.0	18.20	3.20
ISLB 500	75.0	95.50	500	180	14.1	9.2	38 579.0	1 063.9	20.10	3.34
ISLB 550	86.3	109.97	550	190	15.0	9.9	53 161.6	1 335.1	21.99	3.48
ISLB 600	99.5	126.69	600	210	15.5	10.5	72 867.6	1 821.9	23.98	3.79
ISMB 100	11.5	14.60	100	75	7.2	4.0	257.5	40.8	4.20	1.67
ISMB 125	13.0	16.60	125	75	7.6	4.4	449.0	43.7	5.20	1.62
ISMB 150	14.9	19.00	150	80	7.6	4.8	726.4	52.6	6.18	1.66
ISMB 175	19.3	24.62	175	90	8.6	5.5	1 272.0	85.0	7.19	1.86
ISMB 200	25.4	32.33	200	100	10.8	5.7	2 235.4	150.0	8.32	2.15
ISMB 225	31.2	39.72	225	110	11.8	6.5	3 441.8	218.3	9.31	2.34
ISMB 250	37.3	47.55	250	125	12.5	6.9	5 131.6	334.5	10.39	2.65
ISMB 300	44.2	56.26	300	140	12.4	7.5	8 603.6	453.9	12.37	2.84
ISMB 350	52.4	66.71	350	140	14.2	8.1	13 630.3	537.7	14.29	2.84
ISMB 400	61.6	78.46	400	140	16.0	8.9	20 458.4	422.1	16.15	2.82
ISMB 450	72.4	92.27	450	150	17.4	9.4	30 390.8	834.0	18.15	3.01
ISMB 500	86.9	110.74	500	180	17.2	10.2	45 218.3	1 369.8	20.21	3.52

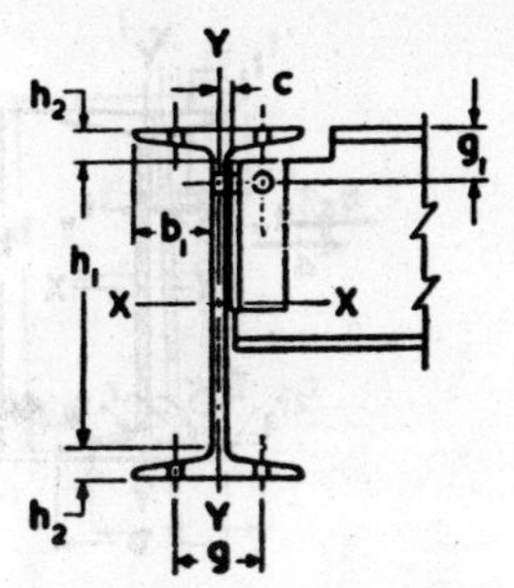

Moduli of Section		Radius at Root	Radius at Toe	Slope of Flange	Connection Details						Maximum Size of Flange Rivet	Designation
Z_{xx}	Z_{yy}	r_1	r_2	D	h_1	h_2	b_1	C	g	g_1 (min)		
cm³	cm³	mm	mm	degrees	mm	mm	mm	mm	mm	mm	mm	
42.9	3.7	5.0	1.5	91.5	130.4	9.80	23.50	3.00	30	45	6	ISJB 150
54.8	3.9	5.0	1.5	91.5	155.0	10.00	23.40	3.10	30	45	6	ISJB 175
78.1	5.8	5.0	1.5	91.5	179.5	10.25	28.38	3.20	30	45	6	ISJB 200
116.3	10.1	6.5	1.5	91.5	201.1	11.95	38.15	3.35	40	45	12	ISJB 225
19.4	4.0	6.5	2.0	91.5	51.7	11.65	23.15	3.35	30	—	6	ISLB 75
33.6	5.1	7.0	3.0	91.5	73.0	13.50	23.00	3.50	30	50	6	ISLB 100
65.1	11.6	8.0	3.0	91.5	95.4	14.80	35.30	3.70	35	50	12	ISLB 125
91.8	13.8	9.5	3.0	91.5	116.9	16.55	37.60	3.90	40	50	12	ISLB 150
125.3	17.7	9.5	3.0	91.5	141.6	16.70	42.45	4.05	50	50	12	ISLB 175
169.7	23.1	9.5	3.0	91.5	165.7	17.15	47.30	4.20	55	50	16	ISLB 200
222.4	22.5	12.0	6.0	98	180.3	22.35	47.10	4.45	55	55	16	ISLB 225
297.4	30.9	13.0	6.5	98	202.6	23.70	59.45	4.55	65	60	22	ISLB 250
392.4	41.0	14.0	7.0	98	223.7	25.65	66.80	4.70	80	60	22	ISLB 275
488.9	50.2	15.0	7.5	98	245.1	27.45	71.65	4.85	90	60	22	ISLB 300
607.7	61.9	16.0	8.0	98	266.5	29.25	79.00	5.00	100	65	25	ISLB 325
751.9	76.6	16.0	8.0	98	288.3	30.85	78.80	5.20	100	65	25	ISLB 350
965.3	86.8	16.0	8.0	98	336.2	31.90	78.50	5.50	100	65	25	ISLB 400
1 223.8	100.4	16.0	8.0	98	384.0	33.00	80.70	5.80	100	70	25	ISLB 450
1 543.2	118.2	17.0	8.5	98	430.2	34.90	85.40	6.10	100	70	28	ISLB 500
1 933.2	140.5	18.0	9.0	98	476.1	36.95	90.05	6.45	100	70	32	ISLB 550
2 428.9	173.5	20.0	10.0	98	520.2	39.90	99.75	6.75	140,100	75	25,32	ISLB 600
51.5	10.9	9.0	4.5	98	65.0	17.50	35.50	3.50	35	55	12	ISMB 100
71.8	11.7	9.0	4.5	98	89.2	17.90	35.30	3.70	35	55	12	ISMB 125
96.9	13.1	9.0	4.5	98	113.9	18.05	37.60	3.90	40	55	12	ISMB 150
145.4	18.9	10.0	5.0	98	134.5	20.25	42.25	4.25	50	55	12	ISMB 175
223.5	30.0	11.0	5.5	98	152.7	23.65	47.15	4.35	55	60	16	ISMB 200
305.9	39.7	12.0	6.0	98	173.3	25.85	51.75	4.75	60	60	20	ISMB 225
410.5	53.5	13.0	6.5	98	194.1	27.95	59.05	4.95	65	65	22	ISMB 250
573.6	64.8	14.0	7.0	98	241.5	29.25	66.25	5.25	80	65	22	ISMB 300
778.9	76.8	14.0	7.0	98	288.0	31.00	65.95	5.55	80	65	22	ISMB 350
1 022.9	88.9	14.0	7.0	98	334.4	32.80	65.55	5.95	80	70	22	ISMB 400
1 350.7	111.2	15.0	7.5	98	379.2	35.40	70.30	6.20	90	70	22	ISMB 450
1 808.7	152.2	17.0	8.5	98	424.1	37.95	84.90	6.60	100	73	28	ISMB 500

(Continued)

TABLE I ROLLED STEEL BEAMS

DIMENSIONS AND PROPERTIES

(Continued)

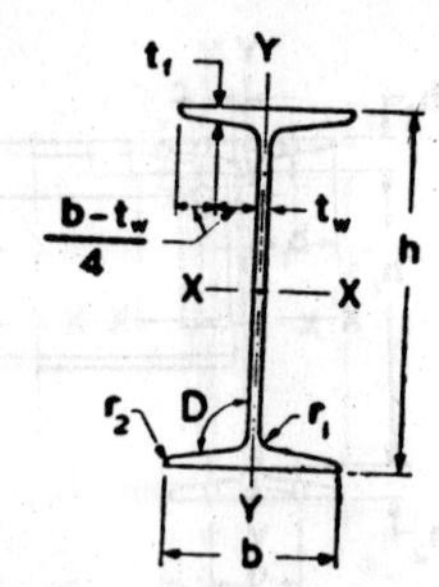

Designation	Weight per Metre	Sectional Area	Depth of Section	Width of Flange	Thick-ness of Flange	Thick-ness of Web	Moments of Inertia		Radii of Gyration	
	w	a	h	b	t_f	t_w	I_{xx}	I_{yy}	r_{xx}	r_{yy}
	kg	cm²	mm	mm	mm	mm	cm⁴	cm⁴	cm	cm
ISMB 550	103.7	132.11	550	190	19.3	11.2	64 893.6	1 833.8	22.16	3.73
ISMB 600	122.6	156.21	600	210	20.8	12.0	91 813.0	2 651.0	24.24	4.12
ISWB 150	17.0	21.67	150	100	7.0	5.4	839.1	94.8	6.22	2.09
ISWB 175	22.1	28.11	175	125	7.4	5.8	1 509.4	188.6	7.33	2.59
ISWB 200	28.8	36.71	200	140	9.0	6.1	2 624.5	328.8	8.46	2.99
ISWB 225	33.9	43.24	225	150	9.9	6.4	3 920.5	448.6	9.52	3.22
ISWB 250	40.9	52.05	250	200	9.0	6.7	5 943.1	857.5	10.69	4.06
ISWB 300	48.1	61.33	300	200	10.0	7.4	9 821.6	990.1	12.66	4.02
ISWB 350	56.9	72.50	350	200	11.4	8.0	15 521.7	1 175.9	14.63	4.03
ISWB 400	66.7	85.01	400	200	13.0	8.6	23 426.7	1 388.0	16.60	4.04
ISWB 450	79.4	101.15	450	200	15.4	9.2	35 057.6	1 706.7	18.63	4.11
ISWB 500	95.2	121.22	500	250	14.7	9.9	52 290.9	2 987.8	20.77	4.96
ISWB 550	112.5	143.34	550	250	17.6	10.5	74 906.1	3 740.6	22.86	5.11
ISWB 600	133.7	170.38	600	250	21.3	11.2	106 198.5	4 702.5	24.97	5.25
ISWB 600	145.1	184.86	600	250	23.6	11.8	115 626.6	5 298.3	25.01	5.35
ISHB 150	27.1	34.48	150	150	9.0	5.4	1 455.6	431.7	6.50	3.54
ISHB 150	30.6	38.98	150	150	9.0	8.4	1 540.0	460.3	6.29	3.44
ISHB 150	34.6	44.08	150	150	9.0	11.8	1 635.6	494.9	6.09	3.35
ISHB 200	37.3	47.54	200	200	9.0	6.1	3 608.4	967.1	8.71	4.51
ISHB 200	40.0	50.94	200	200	9.0	7.8	3 721.8	994.6	8.55	4.42
ISHB 225	43.1	54.94	225	225	9.1	6.5	5 279.5	1 353.8	9.80	4.96
ISHB 225	46.8	59.66	225	225	9.1	8.6	5 478.8	1 396.6	9.58	4.84
ISHB 250	51.0	64.96	250	250	9.7	6.9	7 736.5	1 961.3	10.91	5.49
ISHB 250	54.7	69.71	250	250	9.7	8.8	7 983.9	2 011.7	10.70	5.37
ISHB 300	58.8	74.85	300	250	10.6	7.6	12 545.2	2 193.6	12.95	5.41
ISHB 300	63.0	80.25	300	250	10.6	9.4	12 950.2	2 246.7	12.70	5.29
ISHB 350	67.4	85.91	350	250	11.6	8.3	19 159.7	2 451.4	14.93	5.34
ISHB 350	72.4	92.21	350	250	11.6	10.1	19 802.8	2 510.5	14.65	5.22
ISHB 400	77.4	98.66	400	250	12.7	9.1	28 083.5	2 728.3	16.87	5.26
ISHB 400	82.2	104.66	400	250	12.7	10.6	28 823.5	2 783.0	16.61	5.16
ISHB 450	87.2	111.14	450	250	13.7	9.8	39 210.8	2 985.2	18.78	5.18
ISHB 450	92.5	117.89	450	250	13.7	11.3	40 349.9	3 045.0	18.50	5.08

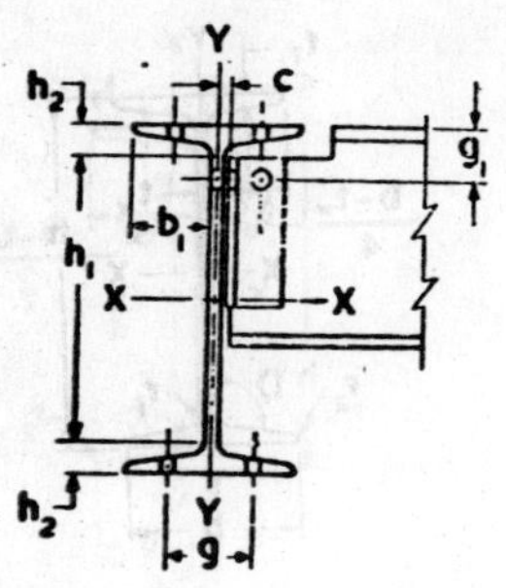

Moduli of Section		Radius at Root	Radius at Toe	Slope of Flange	Connection Details						Maximum Size of Flange Rivet	Designation
Z_{xx}	Z_{yy}	r_1	r_2	D	h_1	h_2	b_1	C	g	g_1 (*min*)		
cm³	cm³	mm	mm	degrees	mm	mm	mm	mm	mm	mm	mm	
2 359.8	193.0	18.0	9.0	98	467.5	41.25	89.40	7.10	100	75	32	ISMB 550
3 060.4	252.5	20.0	10.0	98	509.7	45.15	99.00	7.50	140,100	80	25, 32	ISMB 600
111.9	19.0	8.0	4.0	96	116.6	16.70	47.30	4.20	55	55	16	ISWB 150
172.5	30.2	8.0	4.0	96	139.5	17.75	59.60	4.40	65	55	22	ISWB 175
262.5	47.0	9.0	4.5	96	158.8	20.60	66.95	4.55	80	55	22	ISWB 200
348.5	59.8	9.0	4.5	96	181.4	21.80	71.80	4.70	90	55	22	ISWB 225
475.4	85.7	10.0	5.0	96	203.8	23.10	96.65	4.85	140,100	60	22, 32	ISWB 250
654.8	99.0	11.0	5.5	96	250.1	24.95	96.30	5.20	140,100	60	22, 32	ISWB 300
887.0	117.6	12.0	6.0	96	295.5	27.25	96.00	5.50	140,100	60	22, 32	ISWB 350
1 171.3	138.8	13.0	6.5	96	340.5	29.75	95.70	5.80	140,100	65	22, 32	ISWB 400
1 558.1	170.7	14.0	7.0	96	384.0	33.00	95.40	6.10	140,100	70	22, 32	ISWB 450
2 091.6	239.0	15.0	7.5	96	431.0	34.50	120.05	6.45	140	70	32	ISWB 500
2 723.9	299.2	16.0	8.0	96	473.4	38.30	119.75	6.75	140	75	32	ISWB 550
3 540.0	376.2	17.0	8.5	96	514.2	42.90	119.40	7.10	140	80	32	ISWB 600
3 854.2	423.9	18.0	9.0	96	507.9	46.05	119.10	7.40	140	80	32	ISWB 600
194.1	57.6	8.0	4.0	94	112.0	19.0	72.30	4.20	90	55	22	ISHB 150
205.3	60.2	8.0	4.0	94	112.0	19.0	70.80	5.70	90	55	22	ISHB 150
218.1	63.2	8.0	4.0	94	112.0	19.0	69.10	7.40	90	55	22	ISHB 150
360.8	96.7	9.0	4.5	94	158.4	20.8	96.95	4.55	140,100	55	22, 32	ISHB 200
372.2	98.6	9.0	4.5	94	158.4	20.8	96.10	5.40	140,100	55	22, 32	ISHB 200
469.3	120.3	10.0	5.0	94	180.5	22.2	109.25	4.75	140	55	28	ISHB 225
487.0	123.0	10.0	5.0	94	180.5	22.2	108.20	5.80	140	55	28	ISHB 225
618.9	156.9	10.0	5.0	94	203.5	23.2	121.55	4.95	140	60	32	ISHB 250
638.7	159.7	10.0	5.0	94	203.5	23.2	120.60	5.90	140	60	32	ISHB 250
836.3	175.5	11.0	5.5	94	249.8	25.1	121.20	5.30	140	60	32	ISHB 300
863.3	178.4	11.0	5.5	94	249.8	25.1	120.30	6.20	140	60	32	ISHB 300
1 094.8	196.1	12.0	6.0	94	296.0	27.0	120.85	5.65	140	60	32	ISHB 350
1 131.6	199.4	12.0	6.0	94	296.0	27.0	119.95	6.55	140	60	32	ISHB 350
1 404.2	218.3	14.0	7.0	94	340.1	29.9	120.45	6.05	140	65	32	ISHB 400
1 444.2	221.3	14.0	7.0	94	340.1	29.9	119.70	6.80	140	65	32	ISHB 400
1 742.7	238.8	15.0	7.5	94	386.2	31.9	120.10	6.40	140	65	32	ISHB 450
1 793.3	242.1	15.0	7.5	94	386.2	31.9	119.35	7.15	140	65	32	ISHB 450

TABLE II ROLLED STEEL CHANNELS

DIMENSIONS AND PROPERTIES

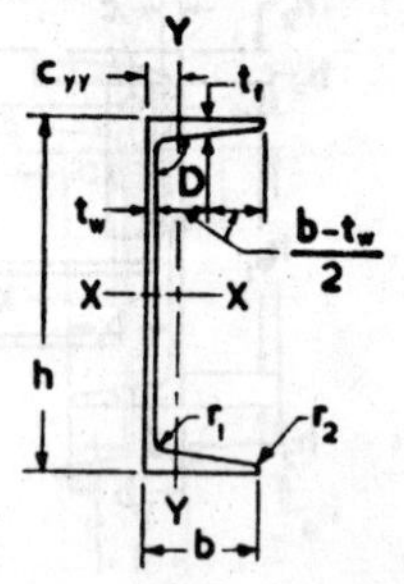

Designation	Weight per Metre	Sectional Area	Depth of Section	Width of Flange	Thickness of Flange	Thickness of Web	Centre of Gravity	Moments of Inertia		Radii of Gyration	
	w	a	h	b	t_f	t_w	c_{yy}	I_{xx}	I_{yy}	r_{xx}	r_{yy}
	kg	cm²	mm	mm	mm	mm	cm	cm⁴	cm⁴	cm	cm
ISJC 100	5.8	7.41	100	45	5.1	3.0	1.40	123.8	14.9	4.09	1.42
ISJC 125	7.9	10.07	125	50	6.6	3.0	1.64	270.0	25.7	5.18	1.60
ISJC 150	9.9	12.65	150	55	6.9	3.6	1.66	471.1	37.9	6.10	1.73
ISJC 175	11.2	14.24	175	60	6.9	3.6	1.75	719.9	50.5	7.11	1.88
ISJC 200	13.9	17.77	200	70	7.1	4.1	1.97	1 161.2	84.2	8.08	2.18
ISLC 75	5.7	7.26	75	40	6.0	3.7	1.35	66.1	11.5	3.02	1.26
ISLC 100	7.9	10.02	100	50	6.4	4.0	1.62	164.7	24.8	4.06	1.57
ISLC 125	10.7	13.67	125	65	6.6	4.4	2.04	356.8	57.2	5.11	2.05
ISLC 150	14.4	18.36	150	75	7.8	4.8	2.38	697.2	103.2	6.16	2.37
ISLC 175	17.6	22.40	175	75	9.5	5.1	2.40	1 148.4	126.5	7.16	2.38
ISLC 200	20.6	26.22	200	75	10.8	5.5	2.35	1 725.5	146.9	8.11	2.37
ISLC 225	24.0	30.53	225	90	10.2	5.8	2.46	2 547.9	209.5	9.14	2.62
ISLC 250	28.0	35.65	250	100	10.7	6.1	2.70	3 687.9	298.4	10.17	2.89
ISLC 300	33.1	42.11	300	100	11.6	6.7	2.55	6 047.9	346.0	11.98	2.87
ISLC 350	38.8	49.47	350	100	12.5	7.4	2.41	9 312.6	394.6	13.72	2.82
ISLC 400	45.7	58.25	400	100	14.0	8.0	2.36	13 989.5	460.4	15.50	2.81
ISMC 75	6.8	8.67	75	40	7.3	4.4	1.31	76.0	12.6	2.96	1.21
ISMC 100	9.2	11.70	100	50	7.5	4.7	1.53	186.7	25.9	4.00	1.49
ISMC 125	12.7	16.19	125	65	8.1	5.0	1.94	416.4	59.9	5.07	1.92
ISMC 150	16.4	20.88	150	75	9.0	5.4	2.22	779.4	102.3	6.11	2.21
ISMC 175	19.1	24.38	175	75	10.2	5.7	2.20	1 223.3	121.0	7.08	2.23
ISMC 200	22.1	28.21	200	75	11.4	6.1	2.17	1 819.3	140.4	8.03	2.23
ISMC 225	25.9	33.01	225	80	12.4	6.4	2.30	2 694.6	187.2	9.03	2.38
ISMC 250	30.4	38.67	250	80	14.1	7.1	2.30	3 816.8	219.1	9.94	2.38
ISMC 300	35.8	45.64	300	90	13.6	7.6	2.36	6 362.6	310.8	11.81	2.61
ISMC 350	42.1	53.66	350	100	13.5	8.1	2.44	10 008.0	430.6	13.66	2.83
ISMC 400	49.4	62.93	400	100	15.3	8.6	2.42	15 082.8	504.8	15.48	2.83

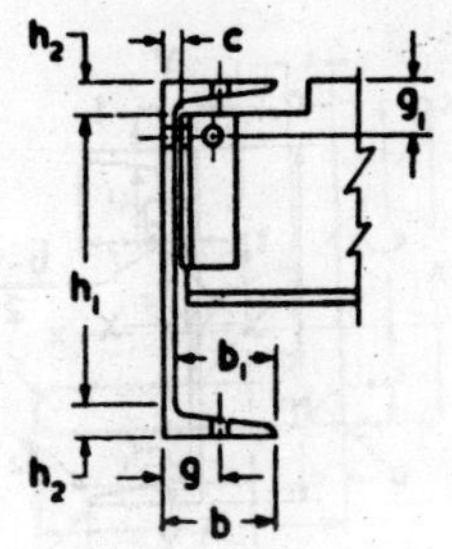

Moduli of Section		Radius at Root	Radius at Toe	Slope of Flange	Connection Details						Maximum Size of Flange Rivet	Designation
Z_{xx}	Z_{yy}	r_1	r_2	D	h_1	h_2	$\frac{b_1}{2}$	C	g	g_1 (*min*)		
cm^3	cm^3	mm	mm	degrees	mm	mm	mm	mm	mm	mm	mm	
24.8	4.8	6.0	2.0	91.5	77.0	11.5	21.0	4.5	25	50	12	ISJC 100
43.2	7.6	6.0	2.5	91.5	98.9	13.1	23.5	4.5	28	50	16	ISJC 125
62.8	9.9	7.0	3.0	91.5	121.2	14.4	25.7	5.1	30	50	20	ISJC 150
82.3	11.9	7.0	3.0	91.5	146.1	14.5	28.2	5.1	35	50	20	ISJC 175
116.1	16.7	8.0	3.5	91.5	168.5	15.8	33.0	5.6	40	50	22	ISJC 200
17.6	4.3	6.0	2.0	91.5	50.4	12.3	18.2	5.2	21	—	12	ISLC 75
32.9	7.3	6.0	2.0	91.5	74.3	12.8	23.0	5.5	28	50	16	ISLC 100
57.1	12.8	7.0	2.5	91.5	96.6	14.2	30.3	5.9	35	50	22	ISLC 125
93.0	20.2	8.0	3.5	91.5	117.0	16.5	35.1	6.3	40	50	25	ISLC 150
131.3	24.8	8.0	4.0	91.5	138.6	18.2	35.0	6.6	40	55	25	ISLC 157
172.6	28.5	8.5	4.5	91.5	160.0	20.0	34.8	7.0	40	55	25	ISLC 200
226.5	32.0	11.0	5.5	96	175.9	24.5	42.1	7.3	50	60	28	ISLC 225
295.0	40.9	11.0	5.5	96	198.9	25.5	47.0	7.6	60	60	28	ISLC 250
403.2	46.4	12.0	6.0	96	245.4	27.3	46.7	8.2	60	60	28	ISLC 300
532.1	52.0	13.0	6.0	96	291.9	29.1	46.3	8.9	60	65	28	ISLC 350
699.5	60.2	14.0	7.0	96	337.1	31.4	46.0	9.5	60	65	28	ISLC 400
20.3	4.7	8.5	4.5	96	41.4	16.8	17.8	5.9	21	—	12	ISMC 75
37.3	7.5	9.0	4.5	96	64.0	18.0	22.7	6.2	28	50	16	ISMC 100
66.6	13.1	9.5	5.0	96	85.4	19.8	30.0	6.5	35	55	22	ISMC 125
103.9	19.4	10.0	5.0	96	106.7	21.7	34.8	6.9	40	55	25	ISMC 150
139.8	22.8	10.5	5.5	96	128.4	23.3	34.7	7.2	40	55	25	ISMC 175
181.9	26.3	11.0	5.5	96	150.2	24.9	34.5	7.6	40	60	25	ISMC 200
239.5	32.8	12.0	6.0	96	170.9	27.1	36.8	7.9	45	60	25	ISMC 225
305.3	38.4	12.0	6.0	96	192.5	28.7	36.5	8.6	45	65	25	ISMC 250
424.2	46.8	13.0	6.5	96	240.7	29.6	41.2	9.1	50	65	28	ISMC 300
571.9	57.0	14.0	7.0	96	288.1	30.9	46.0	9.6	60	65	28	ISMC 350
754.1	66.6	15.0	7.5	96	332.8	33.6	45.7	10.1	60	70	28	ISMC 400

TABLE III ROLLED STEEL EQUAL ANGLES

DIMENSIONS AND PROPERTIES

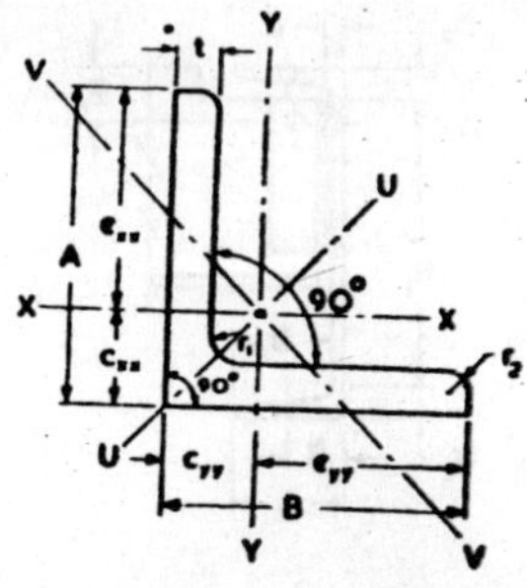

Designation	Size	Thickness	Sectional Area	Weight per Metre	Centre of Gravity	Distance of Extreme Fibre
	$A \times B$	t	a	w	$C_{xx}=C_{yy}$	$e_{xx}=e_{yy}$
	mm mm	mm	cm²	kg	cm	cm
ISA 2020	20×20	3.0	1.12	0.9	0.59	1.41
		4.0	1.45	1.1	0.63	1.37
ISA 2525	25×25	3.0	1.41	1.1	0.71	1.79
		4.0	1.84	1.4	0.75	1.75
		5.0	2.25	1.8	0.79	1.71
ISA 3030	30×30	3.0	1.73	1.4	0.83	2.17
		4.0	2.26	1.8	0.87	2.13
		5.0	2.77	2.2	0.92	2.08
ISA 3535	35×35	3.0	2.03	1.6	0.95	2.55
		4.0	2.66	2.1	1.00	2.50
		5.0	3.27	2.6	1.04	2.46
		6.0	3.86	3.0	1.08	2.42
ISA 4040	40×40	3.0	2.34	1.8	1.08	2.92
		4.0	3.07	2.4	1.12	2.88
		5.0	3.78	3.0	1.16	2.84
		6.0	4.47	3.5	1.20	2.80
ISA 4545	45×45	3.0	2.64	2.1	1.20	3.30
		4.0	3.47	2.7	1.25	3.25
		5.0	4.28	3.4	1.29	3.21
		6.0	5.07	4.0	1.33	3.17
ISA 5050	50×50	3.0	2.95	2.3	1.32	3.68
		4.0	3.88	3.0	1.37	3.63
		5.0	4.79	3.8	1.41	3.59
		6.0	5.68	4.5	1.45	3.55
ISA 5555	55×55	5.0	5.27	4.1	1.53	3.97
		6.0	6.26	4.9	1.57	3.93
		8.0	8.18	6.4	1.65	3.85
		10.0	10.02	7.9	1.72	3.78
ISA 6060	60×60	5.0	5.75	4.5	1.65	4.35
		6.0	6.84	5.4	1.69	4.31
		8.0	8.96	7.0	1.77	4.23
		10.0	11.00	8.6	1.85	4.15
ISA 6565	65×65	5.0	6.25	4.9	1.77	4.73
		6.0	7.44	5.8	1.81	4.69
		8.0	9.76	7.7	1.89	4.61
		10.0	12.00	9.4	1.97	4.53

Moments of Inertia			Radii of Gyration			Moduli of Section	Radius at Root	Radius at Toe	Product of Inertia	Designation
$I_{xx}=I_{yy}$	I_{uu}	I_{vv}	$r_{xx}=r_{yy}$	r_{uu}	r_{vv}	$Z_{xx}=Z_{yy}$	r_1	r_2	I_{xy}	
cm⁴	cm⁴	cm⁴	cm	cm	cm	cm³	mm	mm	cm⁴	
0.4	0.6	0.2	0.58	0.73	0.37	0.3	4.0	2.5	0.2	ISA 2020
0.5	0.8	0.2	0.58	0.72	0.37	0.4			0.3	
0.8	1.2	0.3	0.73	0.93	0.47	0.4	4.5	3.0	0.4	ISA 2525
1.0	1.6	0.4	0.73	0.91	0.47	0.6			0.6	
1.2	1.8	0.5	0.72	0.91	0.47	0.7			0.7	
1.4	2.2	0.6	0.89	1.13	0.57	0.6	5.0	3.0	0.8	ISA 3030
1.8	2.8	0.7	0.89	1.12	0.57	0.8			1.0	
2.1	3.4	0.9	0.88	1.11	0.57	1.0			1.2	
2.3	3.6	0.9	1.05	1.33	0.67	0.9	5.0	3.0	1.3	ISA 3535
2.9	4.7	1.2	1.05	1.32	0.67	1.2			1.7	
3.5	5.6	1.5	1.04	1.31	0.67	1.4			2.1	
4.1	6.5	1.7	1.03	1.29	0.67	1.7			2.4	
3.4	5.5	1.4	1.21	1.54	0.77	1.2	5.5	3.0	2.0	ISA 4040
4.5	7.1	1.8	1.21	1.53	0.77	1.6			2.6	
5.4	8.6	2.2	1.20	1.51	0.77	1.9			3.2	
6.3	10.0	2.6	1.19	1.50	0.77	2.3			3.7	
5.0	8.0	2.0	1.38	1.74	0.87	1.5	5.5	3.0	2.9	ISA 4545
6.5	10.4	2.6	1.37	1.73	0.87	2.0			3.8	
7.9	12.6	3.2	1.36	1.72	0.87	2.5			4.6	
9.2	14.6	3.8	1.35	1.70	0.87	2.9			5.4	
6.9	11.1	2.8	1.53	1.94	0.97	1.9	6.0	3.0	4.1	ISA 5050
9.1	14.5	3.6	1.53	1.93	0.97	2.5			5.3	
11.0	17.6	4.5	1.52	1.92	0.97	3.1			6.5	
12.9	20.6	5.3	1.51	1.90	0.96	3.6			7.6	
14.7	23.5	5.9	1.67	2.11	1.06	3.7	6.5	4.0	8.6	ISA 5555
17.3	27.5	7.0	1.66	2.10	1.06	4.4			10.1	
22.0	34.9	9.1	1.64	2.07	1.06	5.7			12.8	
26.3	41.5	11.2	1.62	2.03	1.06	7.0			15.1	
19.2	30.6	7.7	1.82	2.31	1.16	4.4	6.5	4.5	11.3	ISA 6060
22.6	36.0	9.1	1.82	2.29	1.15	5.2			13.3	
29.0	46.0	11.9	1.80	2.27	1.15	6.8			16.9	
34.8	54.9	14.6	1.78	2.23	1.15	8.4			20.1	
24.7	39.4	9.9	1.99	2.51	1.26	5.2	6.5	4.5	14.5	ISA 6565
29.1	46.5	11.7	1.98	2.50	1.26	6.2			17.2	
37.4	59.5	15.3	1.96	2.47	1.25	8.1			22.0	
45.0	71.3	18.8	1.94	2.44	1.25	9.9			26.2	

(Continued)

TABLE III ROLLED STEEL EQUAL ANGLES

DIMENSIONS AND PROPERTIES

(Continued)

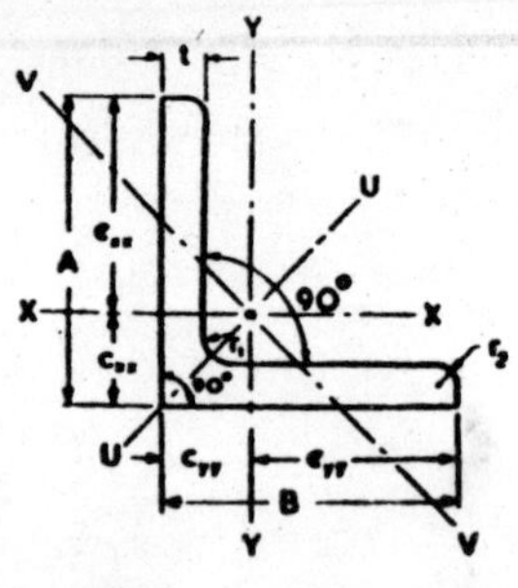

Designation	Size $A \times B$	Thickness t	Sectional Area a	Weight per Metre w	Centre of Gravity $C_{xx}=C_{yy}$	Distance of Extreme Fibre $e_{xx}=e_{yy}$
	mm mm	mm	cm^2	kg	cm	cm
ISA 7070	70×70	5.0	6.77	5.3	1.89	5.11
		6.0	8.06	6.3	1.94	5.06
		8.0	10.58	8.3	2.02	4.98
		10.0	13.02	10.2	2.10	4.90
ISA 7575	75×75	5.0	7.27	5.7	2.02	5.48
		6.0	8.66	6.8	2.06	5.44
		8.0	11.38	8.9	2.14	5.36
		10.0	14.02	11.0	2.22	5.28
ISA 8080	80×80	6.0	9.29	7.3	2.18	5.82
		8.0	12.21	9.6	2.27	5.73
		10.0	15.05	11.8	2.34	5.66
		12.0	17.81	14.0	2.42	5.58
ISA 9090	90×90	6.0	10.47	8.2	2.42	6.58
		8.0	13.79	10.8	2.51	6.49
		10.0	17.03	13.4	2.59	6.41
		12.0	20.19	15.8	2.66	6.34
ISA 100100	100×100	6.0	11.67	9.2	2.67	7.33
		8.0	15.39	12.1	2.76	7.24
		10.0	19.03	14.9	2.84	7.16
		12.0	22.59	17.7	2.92	7.08
ISA 110110	110×110	8.0	17.02	13.4	3.00	8.00
		10.0	21.06	16.5	3.08	7.92
		12.0	25.02	19.6	3.16	7.84
		15.0	30.81	24.2	3.27	7.73
ISA 130130	130×130	8.0	20.22	15.9	3.50	9.50
		10.0	25.06	19.7	3.58	9.42
		12.0	29.82	23.4	3.66	9.34
		15.0	36.81	28.9	3.78	9.22
ISA 150150	150×150	10.0	29.03	22.8	4.06	10.94
		12.0	34.59	27.2	4.14	10.86
		15.0	42.78	33.6	4.26	10.74
		18.0	50.79	39.9	4.38	10.62
ISA 200200	200×200	12.0	46.61	36.6	5.36	14.64
		15.0	57.80	45.4	5.49	14.51
		18.0	68.81	54.0	5.61	14.39
		25.0	93.80	73.6	5.88	14.12

Radii of Gyration				Moduli of Section		tan α	Radius at Root	Radius at Toe	Product of Inertia	Designation
r_{xx}	r_{yy}	r_{uu}	r_{vv}	Z_{xx}	Z_{yy}		r_1	r_2	I_{yy}	
cm	cm	cm	cm	cm^3	cm^3		mm	mm	cm^4	
0.92	0.54	0.99	0.41	0.6	0.3	0.43	4.5	3.0	0.4	ISA 3020
0.92	0.54	0.98	0.41	0.8	0.4	0.42			0.5	
0.91	0.53	0.97	0.41	1.0	0.4	0.41			0.6	
1.25	0.68	1.33	0.52	1.1	0.5	0.38	5.0	3.0	0.9	ISA 4025
1.25	0.68	1.32	0.52	1.4	0.6	0.38			1.2	
1.24	0.67	1.31	0.52	1.8	0.7	0.37			1.4	
1.23	0.66	1.29	0.52	2.1	0.9	0.37			1.6	
1.42	0.84	1.52	0.63	1.4	0.7	0.44	5.0	3.0	1.5	ISA 4530
1.41	0.84	1.51	0.63	1.9	0.9	0.43			1.9	
1.40	0.83	1.50	0.63	2.3	1.1	0.43			2.3	
1.39	0.82	1.49	0.63	2.7	1.3	0.42			2.7	
1.59	0.82	1.67	0.65	1.7	0.7	0.36	5.5	3.0	1.7	ISA 5030
1.58	0.82	1.66	0.63	2.3	0.9	0.36			2.3	
1.57	0.81	1.65	0.63	2.8	1.1	0.35			2.7	
1.56	0.80	1.64	0.63	3.4	1.3	0.35			3.1	
1.89	1.12	2.02	0.85	4.2	2.0	0.44	6.0	4.0	5.8	ISA 6040
1.88	1.11	2.01	0.85	5.0	2.3	0.43			6.8	
1.86	1.10	1.98	0.84	6.5	3.0	0.42			8.5	
2.05	1.28	2.22	0.96	5.0	2.5	0.47	6.0	4.0	8.0	ISA 6545
2.04	1.27	2.21	0.95	5.9	3.0	0.47			9.4	
2.02	1.25	2.18	0.95	7.7	3.9	0.46			11.8	
2.22	1.26	2.36	0.96	5.7	2.5	0.41	6.5	4.0	8.9	ISA 7045
2.21	1.25	2.35	0.96	6.8	3.0	0.41			10.5	
2.19	1.24	2.32	0.95	8.9	3.9	0.40			13.2	
2.16	1.22	2.29	0.95	10.9	4.8	0.39			15.5	
2.38	1.42	2.56	1.07	6.7	3.2	0.44	6.5	4.0	11.8	ISA 7550
2.37	1.41	2.55	1.07	8.0	3.8	0.44			13.9	
2.35	1.40	2.52	1.06	10.4	4.9	0.43			17.7	
2.33	1.38	2.49	1.06	12.7	6.0	0.42			20.9	
2.55	1.40	2.70	1.07	7.5	3.2	0.39	7.0	4.5	12.9	ISA 8050
2.54	1.39	2.69	1.07	9.0	3.8	0.39			15.2	
2.52	1.37	2.66	1.06	11.7	4.9	0.38			19.3	
2.49	1.36	2.63	1.06	[illegible]	6.0	0.38			22.9	

(Continued)

TABLE IV ROLLED STEEL UNEQUAL ANGLES

DIMENSIONS AND PROPERTIES

(Continued)

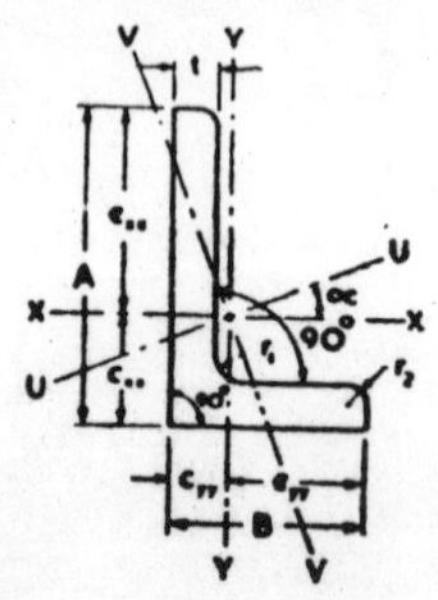

Designation	Size	Thickness	Sectional Area	Weight per Metre	Centre of Gravity		Distance of Extreme Fibre		Moments of Inertia			
	$A \times B$	t	a	w	C_{xx}	C_{yy}	e_{xx}	e_{yy}	I_{xx}	I_{yy}	I_{uu}	I_{vv}
	mm mm	mm	cm²	kg	cm	cm	cm	cm	cm⁴	cm⁴	cm⁴	cm⁴
ISA 9060	90×60	6.0	8.65	6.8	2.87	1.39	6.13	4.61	70.6	25.2	81.5	14.3
		8.0	11.37	8.9	2.96	1.48	6.04	4.52	91.5	32.4	105.3	18.6
		10.0	14.01	11.0	3.04	1.55	5.96	4.45	110.9	39.1	127.3	22.8
		12.0	16.57	13.0	3.12	1.63	5.88	4.37	129.1	45.2	147.5	26.8
ISA 10065	100×65	6.0	9.55	7.5	3.19	1.47	6.81	5.03	96.7	32.4	110.6	18.6
		8.0	12.57	9.9	3.28	1.55	6.72	4.93	125.9	41.9	143.6	24.2
		10.0	15.51	12.2	3.37	1.63	6.63	4.87	153.2	50.7	174.2	29.7
ISA 10075	100×75	6.0	10.14	8.0	3.01	1.78	6.99	5.72	100.9	48.7	124.0	25.6
		8.0	13.36	10.5	3.10	1.87	6.90	5.63	131.6	63.3	161.3	33.6
		10.0	16.50	13.0	3.19	1.95	6.81	5.55	160.4	76.9	196.1	41.2
		12.0	19.56	15.4	3.27	2.03	6.73	5.47	187.5	89.5	228.4	48.6
ISA 12575	125×75	6.0	11.66	9.2	4.05	1.59	8.45	5.91	187.8	51.6	208.9	30.5
		8.0	15.38	12.1	4.15	1.68	8.35	5.82	245.5	67.2	272.8	40.0
		10.0	19.02	14.9	4.24	1.76	8.26	5.74	300.3	81.6	332.9	49.1
ISA 12595	125×95	6.0	12.86	10.1	3.70	2.22	8.80	7.28	203.2	102.1	252.3	52.9
		8.0	16.98	13.3	3.80	2.31	8.70	7.19	266.0	133.3	329.7	69.6
		10.0	21.02	16.5	3.88	2.39	8.62	7.11	325.8	162.7	402.9	85.6
		12.0	24.98	19.6	3.96	2.47	8.54	7.03	382.6	190.4	472.0	101.0
ISA 15075	150×75	8.0	17.42	13.7	5.23	1.53	9.77	5.97	407.2	70.2	432.8	44.5
		10.0	21.56	16.9	5.32	1.61	9.68	5.89	499.1	85.3	529.8	54.6
		12.0	25.62	20.1	5.41	1.69	9.59	5.81	587.0	99.5	622.2	64.3
ISA 150115	150×115	8.0	20.58	16.2	4.46	2.73	10.54	8.77	465.7	238.9	581.2	123.3
		10.0	25.52	20.0	4.55	2.82	10.45	8.68	573.3	293.4	714.3	152.4
		12.0	30.38	23.8	4.64	2.90	10.36	8.60	676.5	345.3	841.4	180.4
		15.0	37.52	29.5	4.76	3.02	10.24	8.48	823.5	418.6	1 020.9	221.2
ISA 200100	200×100	10.0	29.03	22.8	6.96	2.01	13.04	7.99	1 2[illegible].0	209.2	1 286.7	132.5
		12.0	34.59	27.2	7.05	2.10	12.95	7.90	1 4[illegible].7	246.2	1 521.0	156.8
		15.0	42.78	33.6	7.18	2.22	12.82	7.78	1 750.5	298.1	1 856.7	191.9
◄ ISA 200150	200×150	10.0	34.00	26.7	5.99	3.51	14.01	11.49	1 377.9	669.6	1 696.[illegible]	350.8
		12.0	40.56	31.8	6.08	3.60	13.92	11.40	1 634.9	793.2	2 010.[illegible]	417.2
		15.0	50.25	39.4	6.20	3.72	13.80	11.28	2 [illegible]03.6	969.9	2 46[illegible]	513.6
		18.0	59.76	46.9	6.33	3.84	13.67	11.16	[illegible] 3[illegible]9.4	1 136.9	2 889.5	[illegible]

Radii of Gyration				Moduli of Section		tan α	Radius at Root	Radius at Toe	Product of Inertia	Designation
r_{xx}	r_{yy}	r_{uu}	r_{vv}	Z_{xx}	Z_{yy}		r_1	r_2	I_{xy}	
cm	cm	cm	cm	cm^3	cm^3		mm	mm	cm^4	
2.86	1.71	3.07	1.28	11.5	5.5	0.44	7.5	5.0	24.5	ISA 9060
2.84	1.69	3.04	1.28	15.1	7.2	0.44			31.5	
2.81	1.67	3.01	1.27	18.6	8.8	0.43			37.8	
2.79	1.65	2.98	1.27	22.0	10.3	0.42			43.3	
3.18	1.84	3.40	1.39	14.2	6.4	0.42	8.0	5.5	32.5	ISA 10065
3.16	1.83	3.38	1.39	18.7	8.5	0.42			42.0	
3.14	1.81	3.35	1.38	23.1	10.4	0.41			50.7	
3.15	2.19	3.50	1.59	14.4	8.5	0.55	8.5	6.0	41.0	ISA 10075
3.14	2.18	3.48	1.59	19.1	11.2	0.55			53.4	
3.12	2.16	3.45	1.58	23.6	13.8	0.55			64.7	
3.10	2.14	3.42	1.58	27.9	16.3	0.54			74.9	
4.01	2.10	4.23	1.62	22.2	8.7	0.37	9.0	6.0	56.7	ISA 12575
4.00	2.09	4.21	1.61	29.4	11.5	0.36			74.0	
3.97	2.07	4.18	1.61	36.3	14.2	0.36			89.9	
3.97	2.82	4.43	2.03	23.1	14.0	0.57	9.0	6.0	84.5	ISA 12595
3.96	2.80	4.41	2.02	30.6	18.5	0.57			110.6	
3.94	2.78	4.38	2.02	37.8	22.9	0.57			135.0	
3.91	2.76	4.35	2.01	44.8	27.1	0.56			157.7	
4.83	2.01	4.98	1.60	41.7	11.8	0.27	10.0	6.0	95.5	ISA 15075
4.81	1.99	4.96	1.59	51.6	14.5	0.26			116.2	
4.79	1.97	4.93	1.58	61.2	17.1	0.26			135.2	
4.76	3.41	5.31	2.45	44.2	27.2	0.58	11.0	7.5	195.9	ISA 150115
4.74	3.39	5.29	2.44	54.9	33.8	0.58			241.0	
4.72	3.37	5.26	2.44	65.3	40.2	0.58			283.6	
4.69	3.34	5.22	2.43	80.4	49.4	0.57			342.8	
6.46	2.68	6.66	2.14	92.8	26.2	0.27	12.0	8.0	284.8	ISA 200100
6.43	2.67	6.63	2.13	110.6	31.1	0.26			335.3	
6.40	2.64	6.59	2.12	136.5	38.3	0.26			405.4	
6.37	4.44	7.06	3.21	98.3	58.3	0.56	13.5	9.5	564.1	ISA 200150
6.35	4.42	7.04	3.21	117.4	69.6	0.56			669.1	
6.32	4.39	7.00	3.20	145.4	86.0	0.55			818.5	
6.28	4.36	6.95	3.19	172.5	101.9				958.1	

TABLE V ROLLED STEEL TEE BARS

DIMENSIONS AND PROPERTIES

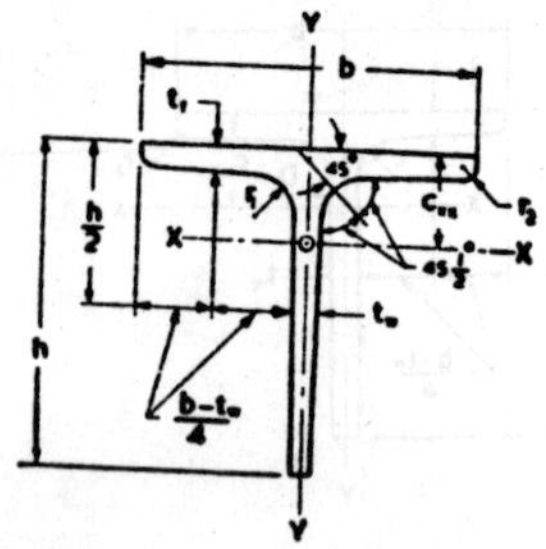

Designation	Weight per Metre w	Sectional Area a	Depth of Section h	Width of Flange b	Thickness of Flange t_f	Thickness of Web t_w	Centre of Gravity C_{xx}	Moments of Inertia I_{xx}	I_{yy}
	kg	cm²	mm	mm	mm	mm	cm	cm⁴	cm⁴
ISNT 20	0.9	1.13	20	20	3.0	3.0	0.60	0.4	0.2
ISNT 30	1.4	1.75	30	30	3.0	3.0	0.83	1.4	0.6
ISNT 40	3.5	4.48	40	40	6.0	6.0	1.20	6.3	3.0
ISNT 50	4.5	5.70	50	50	6.0	6.0	1.44	12.7	5.9
ISNT 60	5.4	6.90	60	60	6.0	6.0	1.67	22.5	10.1
ISNT 80	9.6	12.25	80	80	8.0	8.0	2.23	71.2	32.3
ISNT 100	15.0	19.10	100	100	10.0	10.0	2.79	173.8	79.9
ISNT 150	22.8	29.08	150	150	10.0	10.0	3.95	603.8	267.5
ISHT 75	15.3	19.49	75	150	9.0	8.4	1.62	96.2	230.2
ISHT 100	20.0	25.47	100	200	9.0	7.8	1.91	193.8	497.3
ISHT 125	27.4	34.85	125	250	9.7	8.8	2.37	415.4	1 005.8
ISHT 150	29.4	37.42	150	250	10.6	7.6	2.66	573.7	1 096.8
ISST 100	8.1	10.37	100	50	10.0	5.8	3.03	99.0	9.6
ISST 150	15.7	19.96	150	75	11.6	8.0	4.75	450.2	37.0
ISST 200	28.4	36.22	200	165	12.5	8.0	4.78	1 267.8	358.2
ISST 250	37.5	47.75	250	180	14.1	9.2	6.40	2 774.4	532.0
ISLT 50	4.0	5.11	50	50	6.4	4.0	1.19	9.9	6.4
ISLT 75	7.1	9.04	75	80	6.8	4.8	1.72	41.9	27.6
ISLT 100	12.7	16.16	100	100	10.8	5.7	2.13	116.6	75.0
ISJT 75	3.5	4.50	75	50	4.6	3.0	2.00	24.8	4.6
ISJT 87.5	4.0	5.14	87.5	50	4.8	3.2	2.50	39.0	4.8
ISJT 100	5.0	6.32	100	60	5.0	3.4	2.81	63.5	8.6
ISJT 112.5	6.4	8.14	112.5	80	5.0	3.7	3.01	101.6	20.2

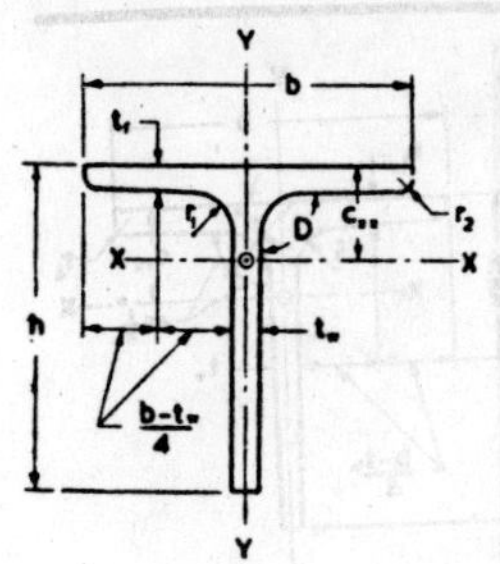

Radii of Gyration		Moduli of Section		Radius at Root	Radius at Toe	Slope of Flange	Designation
r_{xx}	r_{yy}	Z_{xx}	Z_{yy}	r_1	r_2	D	
cm	cm	cm^3	cm^3	mm	mm	degrees	
0.59	0.39	0.3	0.2	4.0	3.0	91	ISNT 20
0.89	0.57	0.6	0.4	5.0	3.5	91	ISNT 30
1.18	0.82	2.2	1.5	5.5	4.0	91	ISNT 40
1.50	1.02	3.6	2.4	6.0	4.0	91	ISNT 50
1.81	1.21	5.2	3.4	6.5	4.5	91	ISNT 60
2.41	1.62	12.3	8.1	8.0	5.5	91	ISNT 80
3.02	2.05	24.1	16.0	9.0	6.0	91	ISNT 100
4.56	3.03	54.6	35.7	10.0	7.0	91	ISNT 150
2.22	3.44	16.4	30.1	8.0	4.0	94	ISHT 75
2.76	4.42	24.0	49.3	9.0	4.5	94	ISHT 100
3.45	5.37	41.0	79.9	10.0	5.0	94	ISHT 125
3.92	5.41	46.5	87.7	11.0	5.5	94	ISHT 150
3.09	0.96	14.2	3.8	8.0	4.0	98	ISST 100
4.75	1.36	43.9	9.9	9.0	4.5	98	ISST 150
5.92	3.15	83.3	43.4	16.0	8.0	98	ISST 200
7.62	3.34	149.2	59.1	17.0	8.5	98	ISST 250
1.39	1.12	2.6	2.5	7.0	3.0	91.5	ISLT 50
2.15	1.75	7.2	6.9	9.5	3.0	91.5	ISLT 75
2.69	2.15	14.8	15.0	11.0	5.5	98	ISLT 100
2.35	1.01	4.5	1.8	5.0	1.5	91.5	ISJT 75
2.75	0.97	6.2	1.9	5.0	1.5	91.5	ISJT 87.5
3.17	1.17	8.8	2.9	5.0	1.5	91.5	ISJT 100
3.53	1.50	12.3	5.1	6.5	1.5	91.5	ISJT 112.5

TABLE I

AREAS OF M.S. REINFORCEMENTS (Round Bars) IN SQUARE CENTIMETRE

Dia of bars \ Number of bars	1	2	3	4	5	6	7	8	9	10	11	12	13	14	15	16	17	18
5	0.196	0.393	0.589	0.785	0.982	1.178	1.374	1.570	1.767	1.963	2.160	2.356	2.552	2.749	2.945	3.142	3.337	3.534
6	0.283	0.566	0.848	1.131	1.414	1.697	1.980	2.262	2.545	2.828	3.111	3.394	3.676	3.959	4.241	4.525	4.806	5.090
7	0.385	0.770	1.155	1.540	1.925	2.309	2.694	3.079	3.464	3.849	4.234	4.620	5.002	5.389	5.773	6.158	6.542	6.928
8	0.503	1.005	1.508	2.011	2.514	3.016	3.519	4.022	4.524	5.027	5.530	6.032	6.534	7.038	7.541	8.043	8.545	9.049
10	0.785	1.571	2.356	3.142	3.927	4.712	5.498	6.283	7.069	7.854	8.640	9.425	10.210	10.996	11.790	12.566	13.350	14.137
12	1.131	2.262	3.393	4.524	5.655	6.786	7.917	9.048	10.179	11.310	12.450	13.572	14.700	15.834	16.960	18.096	19.230	20.358
14	1.539	3.078	4.617	6.156	7.695	9.234	10.773	12.312	13.851	15.390	16.930	18.468	20.010	21.546	23.100	24.624	26.170	27.702
16	2.010	4.020	6.030	8.040	10.050	12.060	14.070	16.080	18.090	20.100	22.120	24.120	26.130	28.140	30.140	32.160	34.180	36.180
18	2.545	5.090	7.635	10.180	12.725	15.270	17.815	20.360	22.905	25.450	27.990	30.540	33.080	35.630	38.170	40.720	43.250	45.810
20	3.142	6.284	9.426	12.568	15.710	18.852	21.994	25.136	28.278	31.420	34.640	37.704	40.840	43.988	47.130	50.272	53.400	56.556
22	3.801	7.602	11.403	15.204	19.005	22.806	26.607	30.408	34.209	38.010	41.810	45.612	49.410	53.214	57.020	60.816	64.620	68.418
24	4.524	9.048	13.572	18.096	22.620	27.144	31.668	36.192	40.716	45.240	49.760	54.288	58.800	63.336	67.850	72.384	76.900	81.432
25	4.909	9.818	14.727	19.636	24.545	29.454	34.363	39.272	44.181	49.090	54.000	58.908	63.810	68.726	73.640	78.544	83.450	88.362
26	5.309	10.618	15.927	21.236	26.545	31.854	37.163	42.472	47.781	53.090	58.390	63.708	69.010	74.326	79.640	84.944	90.240	95.562
28	6.158	12.316	18.474	24.632	30.790	36.948	43.106	49.244	55.422	61.580	67.740	73.896	80.040	86.212	92.370	98.488	104.700	110.844
30	7.068	14.136	21.204	28.272	35.340	42.408	49.476	56.544	63.612	70.680	77.750	84.816	91.870	98.952	106.000	113.088	120.100	127.224
32	8.042	16.084	24.126	32.168	40.210	48.252	56.294	64.336	72.374	80.420	88.470	96.504	104.600	112.588	120.600	128.672	136.800	144.748
34	9.078	18.156	27.234	36.312	45.390	54.468	63.546	72.624	81.702	90.780	99.860	108.936	118.000	127.092	136.100	145.248	154.300	163.404
36	10.180	20.360	30.540	40.720	50.900	61.080	71.260	81.440	91.620	101.800	111.900	122.160	132.300	142.520	152.700	162.880	173.000	183.240
40	12.570	25.140	37.710	50.280	62.850	75.420	87.990	100.560	113.130	125.700	138.200	150.840	163.300	175.980	188.500	201.120	213.600	226.260

TABLE II

WEIGHTS OF M.S. REINFORCEMENTS (Round Bars) IN KILOGRAMS PER METRE LENGTH

Diameter of bars in mm ↓ / Number of bars →	1	2	3	4	5	6	7	8	9	10	11	12	13	14	15	16	17	18
5	0.154	0.308	0.462	0.614	0.770	0.924	1.078	1.232	1.386	1.540	1.694	1.849	2.003	2.157	2.311	2.465	2.619	2.773
6	0.222	0.444	0.666	0.888	1.110	1.331	1.553	1.775	1.997	2.219	2.441	2.663	2.884	3.107	3.329	3.550	3.761	3.994
7	0.302	0.604	0.906	1.208	1.510	1.812	2.114	2.416	2.718	3.020	3.322	3.624	3.925	4.228	4.530	4.832	5.134	5.436
8	0.395	0.789	1.184	1.578	1.973	2.367	2.762	3.156	3.551	3.945	4.339	4.734	5.128	5.523	5.917	6.312	6.705	7.101
10	0.614	1.227	1.841	2.454	3.068	3.682	4.295	4.909	5.522	6.136	6.779	7.363	8.011	8.590	9.245	9.818	10.470	11.045
12	0.888	1.775	2.663	3.550	4.438	5.326	6.213	7.101	7.988	8.876	9.763	10.651	11.530	12.426	13.310	14.202	15.090	15.977
14	1.208	2.416	3.624	4.832	6.040	7.248	8.456	9.664	10.872	12.080	13.290	14.496	15.700	16.912	18.120	19.328	20.530	21.744
16	1.578	3.156	4.734	6.312	7.990	9.468	11.046	12.624	14.202	15.780	17.360	18.936	20.500	22.092	23.670	25.248	26.810	28.404
18	1.996	3.992	5.988	7.984	9.980	11.976	13.972	15.968	17.964	19.960	21.970	23.952	25.950	27.944	29.950	31.936	33.940	35.928
20	2.465	4.930	7.395	9.860	12.325	14.790	17.255	19.720	22.185	24.650	27.120	29.580	32.050	34.510	36.980	39.440	41.910	44.370
22	2.983	5.966	8.949	11.932	14.915	17.898	20.881	23.864	26.847	29.830	32.810	35.796	38.770	41.762	44.740	47.728	50.700	53.694
24	3.550	7.100	10.650	14.200	17.750	21.300	24.850	28.400	31.950	35.500	39.040	42.600	46.140	49.700	53.250	56.800	60.340	63.900
25	3.852	7.704	11.556	15.408	19.260	23.112	26.964	30.816	34.668	38.520	42.370	46.224	50.070	53.928	57.790	61.632	65.480	69.336
26	4.166	8.332	12.498	16.664	20.830	24.996	29.162	33.428	37.494	41.660	45.820	49.992	54.160	58.324	62.480	66.856	70.810	74.988
28	4.832	9.664	14.496	19.328	24.160	28.992	33.824	38.656	43.488	48.320	53.150	57.984	62.810	67.648	72.610	77.312	82.130	86.976
30	5.546	11.092	16.638	22.184	27.730	33.276	38.822	44.368	49.914	55.460	61.010	66.552	72.090	77.644	83.350	88.736	94.280	99.828
32	6.311	12.622	18.933	25.244	31.555	37.866	44.177	50.488	56.799	63.110	69.420	75.732	82.040	88.354	94.840	100.976	107.300	113.598
34	7.124	14.248	21.372	28.496	35.620	42.744	49.868	56.992	64.116	71.240	78.360	85.488	92.600	99.736	107.000	113.984	121.100	128.232
36	7.987	15.974	23.961	31.948	39.935	47.922	55.909	63.896	71.883	79.870	87.860	95.844	103.900	111.818	120.000	127.792	135.800	143.766
40	9.860	19.720	29.580	39.440	49.300	59.160	69.020	78.880	88.740	98.600	108.500	118.320	128.100	138.040	148.200	157.760	167.600	177.480

TABLE III

WEIGHTS OF M. S. REINFORCEMENTS (Round Bars) IN KILOGRAMS PER SQ. METRE FOR VARIOUS SPACINGS OF BARS IN CENTIMETRES

Spacings in cm / Diameter in mm	5	6	7	8	10	12	14	16	18	20	22	24	25
7.0	2.20	3.17	4.31	5.64	8.80	12.68	17.26	22.54	28.52	35.22	42.8[illegible]	50.71	55.04
7.5	2.05	2.96	4.03	5.26	8.22	11.83	16.11	21.03	26.62	32.87	39.7[illegible]	47.33	51.36
8.0	1.93	2.77	3.78	4.93	7.70	11.09	15.10	19.72	24.96	30.82	37.[illegible]	44.37	48.15
8.5	1.81	2.61	3.55	4.64	7.25	10.44	14.21	18.57	23.49	29.00	3[illegible]	41.77	45.32
9.0	1.71	2.47	3.36	4.38	6.85	9.86	13.43	17.53	22.19	27.39	3[illegible].14	39.45	42.81
9.5	1.62	2.34	3.18	4.15	6.49	9.34	12.72	16.61	21.02	25.95	31.39	37.37	40.55
10.0	1.54	2.22	3.02	3.95	6.16	8.88	12.08	15.78	19.96	24.65	2[illegible].83	35.50	38.52
10.5	1.47	2.11	2.88	3.76	5.87	8.45	11.50	15.03	19.01	23.48	28.41	33.81	36.68
11.0	1.40	2.02	2.75	3.59	5.60	8.07	10.98	14.34	18.15	22.42	27.11	32.27	35.01
11.5	1.34	1.93	2.63	3.43	5.36	7.72	10.51	13.72	17.36	21.44	25.93	30.87	33.50
12.0	1.28	1.85	2.52	3.29	5.14	7.40	10.07	13.14	16.63	20.54	24.85	29.58	32.10
12.5	1.23	1.78	2.42	3.16	4.93	7.10	9.67	12.62	15.97	19.72	23.86	28.40	30.82
13.0	1.19	1.71	2.32	3.04	4.74	6.83	9.29	12.13	15.36	18.97	22.95	27.31	29.63
13.5	1.14	1.65	2.24	2.92	4.57	6.58	8.95	11.69	14.79	18.27	22.10	26.29	28.54
14.0	1.10	1.59	2.16	2.82	4.40	6.34	8.63	11.27	14.26	17.61	21.31	25.36	27.52
14.5	1.06	1.53	2.08	2.72	4.15	6.12	8.33	10.87	13.77	16.62	20.57	24.47	26.57
15.0	1.03	1.48	2.01	2.63	4.11	5.92	8.06	10.52	13.31	16.44	19.89	23.67	25.68
15.5	1.00	1.43	1.95	2.55	3.98	5.73	7.80	10.19	12.90	15.92	19.2[illegible]	22.93	24.88
16.0	0.96	1.39	1.89	2.47	3.85	5.55	7.55	9.86	12.48	15.41	18.64	22.19	24.08
16.5	0.93	1.35	1.83	2.39	3.74	5.38	7.33	9.57	12.11	14.96	18.10	21.54	23.37
17.0	0.91	1.31	1.78	2.32	3.63	5.22	7.11	9.29	11.75	14.50	17.55	20.89	22.66
17.5	0.88	1.27	1.73	2.26	3.52	5.08	6.91	9.03	11.42	14.10	17.06	20.31	22.03
18.0	0.86	1.23	1.68	2.19	3.42	4.93	6.72	8.77	11.10	13.70	16.57	19.73	21.41
18.5	0.83	1.20	1.63	2.13	3.33	4.80	6.54	8.54	10.80	13.34	16.13	19.21	20.84
19.0	0.81	1.17	1.59	2.08	3.24	4.67	6.36	8.31	10.51	12.98	15.70	18.69	20.28
19.5	0.79	1.14	1.55	2.02	3.16	4.56	6.20	8.10	10.25	12.65	15.30	18.22	19.77
20.0	0.77	1.11	1.51	1.97	3.08	4.44	6.04	7.89	9.98	12.33	14.92	17.75	19.26
20.5	0.75	1.08	1.47	1.93	3.01	4.33	5.90	7.70	9.74	12.03	14.56	17.33	18.80
21.0	0.73	1.06	1.44	1.88	2.93	4.23	5.75	7.52	9.51	11.74	14.21	16.91	18.34
21.5	0.72	1.03	1.41	1.84	2.87	4.13	5.62	7.34	9.29	11.48	13.88	16.52	17.92
22.0	0.70	1.01	1.37	1.79	2.80	4.03	5.49	7.17	9.08	11.21	13.56	16.14	17.51
22.5	0.69	0.99	1.34	1.75	2.74	3.95	5.37	7.02	8.88	10.97	13.26	15.79	17.13
23.0	0.67	0.97	1.31	1.72	2.68	3.86	5.26	6.86	8.68	10.72	12.97	15.44	16.75
23.5	0.66	0.94	1.29	1.68	2.62	3.78	5.15	6.72	8.50	10.50	12.70	15.11	16.40
24.0	0.64	0.92	1.26	1.64	2.57	3.70	5.04	6.57	8.32	10.27	12.43	14.79	16.05
24.5	0.63	0.91	1.23	1.61	2.52	3.62	4.93	6.44	8.15	10.07	12.18	14.50	15.73
25.0	0.62	0.89	1.21	1.58	2.47	3.55	4.83	6.31	7.99	9.86	11.93	14.20	15.41
25.5	0.60	0.87	1.19	1.55	2.42	3.48	4.74	6.19	7.83	9.67	11.70	13.93	15.11
26.0	0.59	0.85	1.16	1.52	2.37	3.41	4.65	6.07	7.68	9.49	11.48	13.66	14.82
26.5	0.58	0.84	1.14	1.49	2.33	3.35	4.56	5.96	7.54	9.31	11.26	13.40	14.54
27.0	0.57	0.82	1.12	1.46	2.28	3.29	4.48	5.85	7.40	9.14	11.05	13.15	14.27
27.5	0.56	0.81	1.10	1.43	2.24	3.23	4.40	5.74	7.26	8.97	10.85	12.91	14.02
28.0	0.55	0.79	1.08	1.41	2.20	3.17	4.32	5.64	7.13	8.81	10.65	12.68	13.76
28.5	0.54	0.78	1.06	1.38	2.14	3.12	4.24	5.54	7.01	8.56	10.47	12.46	13.52
29.0	0.53	0.77	1.04	1.36	2.08	3.06	4.17	5.44	6.89	8.31	10.29	12.24	13.29
29.5	0.52	0.75	1.02	1.34	2.07	3.01	4.10	5.35	6.77	8.26	10.11	12.03	13.06
30.0	0.51	0.74	1.01	1.31	2.05	2.96	4.03	5.26	6.66	8.22	9.94	11.83	12.84

TABLE IV

AREAS OF M. S. REINFORCEMENTS (Round bars) IN SQUARE CENTIMETRE

per metre width for various spacings in centimetres

Spacings in cm \ Diameter in mm	5	6	7	8	10	12	14	16	18	20	22	24	25
7.0	2[illegible]	4.04	5.50	7.18	11.22	16.15	22.00	28.72	36.35	44.88	54.30	64.63	70.13
7.5	[illegible]62	3.77	5.13	6.70	10.47	15.08	20.52	26.80	33.92	41.89	50.67	60.32	65.45
8.0	[illegible]5	3.53	4.81	6.28	9.82	14.14	19.24	25.13	31.81	39.27	47.51	56.54	61.37
8.5	[illegible]	3.33	4.53	5.91	9.24	13.30	18.11	23.65	29.93	36.97	44.72	53.22	57.76
9.0	2.1[illegible]	3.14	4.28	5.59	8.73	12.57	17.11	22.35	28.28	34.91	42.24	50.27	54.55
9.5	2.07	2.[illegible]	4.05	5.29	8.27	11.91	16.21	21.16	26.79	33.08	40.01	47.62	51.68
10.0	1.9[illegible]	2.83	3.85	5.03	7.85	11.31	15.39	20.10	25.45	31.42	38.04	45.24	49.09
10.5	1.8[illegible]	2.69	3.67	4.79	7.48	10.77	14.67	19.14	24.23	29.92	36.20	43.08	46.76
11.0	1.78	2.57	3.50	4.57	7.14	10.28	14.00	18.28	23.13	28.56	34.55	41.12	44.63
11.5	1.71	2.46	3.35	4.37	6.83	9.84	13.39	17.48	22.13	27.32	33.06	39.33	42.69
12.0	1.64	2.36	3.21	4.19	6.55	9.43	12.83	16.75	21.20	26.18	31.67	37.70	40.91
12.5	1.57	2.26	3.08	4.02	6.28	9.05	12.31	16.08	20.35	25.14	30.41	36.19	39.27
13.0	1.51	2.18	2.96	3.87	6.04	8.70	11.84	15.46	19.57	24.17	29.24	34.80	37.77
13.5	1.45	2.09	2.85	3.72	5.82	8.38	11.40	14.89	18.85	23.28	28.16	33.52	36.37
14.0	[illegible]	2.02	2.75	3.59	5.61	8.08	11.00	14.36	18.18	22.44	27.15	32.32	35.07
14.5	1.35	1.95	2.66	3.47	5.42	7.80	10.62	13.87	17.55	21.67	26.21	31.20	33.86
15.0	1.3[illegible]	1.89	2.57	3.35	5.24	7.54	10.26	13.40	16.96	20.95	25.34	30.16	32.73
15.5	[illegible]7	1.83	2.49	3.24	5.07	7.31	9.94	12.98	16.43	20.29	24.55	29.22	31.71
16.0	1.23	1.77	2.41	3.14	4.91	7.07	9.62	12.57	15.91	19.64	23.76	28.27	30.69
16.5	1.19	1.72	2.33	3.05	4.76	6.86	9.34	12.20	15.44	19.06	23.06	27.44	29.78
17.0	1.16	1.66	2.26	2.96	4.62	6.65	9.06	11.83	14.97	18.49	22.36	26.61	28.88
17.5	1.12	1.62	2.20	2.88	4.49	6.47	8.81	11.50	14.55	17.97	21.74	25.87	28.08
18.0	1.09	1.57	2.14	2.79	4.36	6.29	8.56	11.18	14.14	17.46	21.12	25.14	27.28
18.5	1.06	1.53	2.08	2.72	4.25	6.12	8.33	10.88	13.77	17.00	20.56	24.47	26.56
19.0	1.03	1.49	2.03	2.65	4.13	5.96	8.11	10.58	13.40	16.54	20.01	23.81	25.84
19.5	1.01	1.45	1.98	2.58	4.03	5.81	7.90	10.32	13.06	16.13	19.51	23.22	25.19
20.0	0.98	1.41	1.92	2.51	3.93	5.66	7.70	10.05	12.73	15.71	19.01	22.62	24.55
20.5	0.96	1.38	1.88	2.45	3.83	5.52	7.52	9.81	12.42	15.34	18.55	22.08	24.46
21.0	0.93	1.35	1.83	2.39	3.74	5.39	7.34	9.57	12.12	14.96	18.10	21.54	23.38
21.5	0.91	1.32	1.79	2.34	3.66	5.26	7.17	9.36	11.84	14.62	17.69	21.05	22.85
22.0	0.89	1.29	1.75	2.28	3.57	5.14	7.00	9.14	11.57	14.28	17.28	20.56	22.32
22.5	0.87	1.26	1.71	2.24	3.49	5.03	6.85	8.94	11.32	13.97	16.90	20.11	21.83
23.0	0.85	1.23	1.67	2.19	3.41	4.92	6.70	8.74	11.07	13.66	16.53	19.67	21.35
23.5	0.84	1.20	1.64	2.14	3.34	4.81	6.56	8.56	10.83	13.38	16.18	19.26	20.90
24.0	0.82	1.18	1.60	2.09	3.27	4.71	6.42	8.34	10.60	13.09	15.84	18.85	20.46
24.5	0.80	1.15	1.57	2.05	3.21	4.62	6.29	8.21	10.39	12.83	15.52	18.47	20.05
25.0	0.79	1.13	1.54	2.01	3.14	4.52	6.16	8.04	10.18	12.57	15.21	18.10	19.64
25.5	0.77	1.11	1.51	1.97	3.08	4.44	6.04	7.89	9.98	12.33	14.91	17.75	19.26
26.0	0.76	1.09	1.48	1.93	3.02	4.35	5.92	7.73	9.79	12.09	14.62	17.40	18.89
26.5	0.74	1.07	1.45	1.90	2.97	4.27	5.81	7.59	9.61	11.86	14.35	17.08	18.54
27.0	0.73	1.05	1.43	1.86	2.91	4.19	5.70	7.45	9.43	11.64	14.08	16.76	18.19
27.5	0.71	1.03	1.40	1.83	2.86	4.11	5.60	7.31	9.25	11.43	13.82	16.45	17.85
28.0	0.70	1.01	1.37	1.80	2.81	4.04	5.50	7.18	9.09	11.22	13.58	16.16	17.53
28.5	0.69	0.99	1.35	1.76	2.76	3.97	5.40	7.06	8.93	11.03	13.34	15.88	17.23
29.0	0.68	0.98	1.33	1.73	2.71	3.90	5.31	6.94	8.78	10.84	13.11	15.60	16.93
29.5	0.67	0.96	1.31	1.70	2.66	3.83	5.22	6.82	8.63	10.65	12.88	15.33	16.64
30.0	0.65	0.94	1.28	1.68	2.62	3.77	5.13	6.70	8.48	10.47	12.67	15.08	16.36

TABLE V

PERIMETERS OF ROUND BARS IN CENTIMETRES

Number of bars \ Dia. of bars in mm	5	6	7	8	10	12	14	16	18	20	22	24	25
1	1.57	1.89	2.20	2.51	3.14	3.77	4.40	5.03	5.66	6.29	6.91	7.54	7.86
2	3.14	3.77	4.40	5.03	6.29	7.54	8.80	10.06	11.31	12.57	13.83	15.09	15.71
3	4.71	5.66	6.60	7.54	9.43	11.31	13.20	15.08	16.97	18.89	20.74	22.63	23.57
4	6.28	7.54	8.80	10.06	12.57	15.08	17.60	20.11	22.63	25.14	27.65	30.17	31.42
5	7.89	9.43	11.00	12.57	15.72	18.86	22.00	25.14	28.29	31.43	34.57	37.72	39.28
6	9.43	11.32	13.20	15.08	18.89	22.63	26.39	30.17	33.94	37.71	41.48	45.26	47.14
7	11.00	13.20	15.40	17.60	22.00	26.40	30.79	35.20	39.60	44.00	48.39	52.80	54.99
8	12.57	15.09	17.60	20.11	25.14	30.17	35.19	40.22	45.26	50.28	55.30	60.34	62.85
9	14.14	16.97	19.80	22.63	28.29	33.94	39.59	45.25	50.91	56.57	62.22	67.89	70.70
10	15.71	18.86	22.00	25.14	31.43	37.71	43.99	50.28	56.57	62.85	69.13	75.43	78.56